PENGUIN BOOKS

KEEPING
chickens

KEEPING
chickens

AN AUSTRALIAN GUIDE

PENGUIN BOOKS

PENGUIN BOOKS

UK | USA | Canada | Ireland | Australia
India | New Zealand | South Africa | China

Penguin Books is part of the Penguin Random House group of companies whose addresses can be found at global.penguinrandomhouse.com.

First published by Penguin Group (Australia), 2007

Text and illustrations copyright © Penguin Group (Australia), 2007

The moral right of the author has been asserted

All rights reserved. Without limiting the rights under copyright reserved above, no part of this publication may be reproduced, stored in or introduced into a retrieval system, or transmitted, in any form or by any means (electronic, mechanical, photocopying, recording or otherwise), without the prior written permission of both the copyright owner and the above publisher of this book.

Cover and text design by Claire Tice © Penguin Group (Australia)
Written by Nicolas Brasch
Illustrations by John Burgess
Cover photograph by Adrian Lambert
Typeset in Perpetua by Post Pre-press Group, Brisbane, Queensland
Scanning and separations by Splitting Image P/L, Clayton, Victoria
Printed and bound in Australia by Griffin Press, an accredited ISO AS/NZS 14001 Environmental Management Systems printer.

National Library of Australia
Cataloguing-in-Publication data:

> Keeping chickens.
> Includes index.
> ISBN 978 0 14 300638 1 (pbk.).
> 1. Chickens – Feeds and feeding. 2. Chickens – Housing.

636.5083

penguin.com.au

Contents

Why Keep Chickens? 1

20 Common Questions Answered 7

10 Steps to Keeping Chickens 15

Know Your Chickens 17

The Right Chickens for Me 27

Housing Your Chickens 41

Caring for Your Chickens 73

Breeding Chickens 103

Chickens for Show 115

Resources and Suppliers 120

Glossary 131

Acknowledgements 133

Index 134

Why Keep Chickens?

THE LURE OF THE COUNTRY

The lure of the country environment can be irresistible. That's why farmstays and B&Bs are booming. City folk just love to get away. And one of the great appeals of country life — apart from the unspoilt landscape, the fresh air, the peace and quiet, and the hospitality — is the sight and sounds (and smells?) of the farm animals. Yet the economic advantages that come from living and working in the major cities are very hard to resist. Few people have the financial freedom and inclination to swap their life in the city for a full-time 'tree change'.

Despite the continuing need and desire for large office blocks and apartment buildings, the squeezing of the 'quarter-acre block' and the construction of enormous shopping centres that even have their own postcodes, there are still opportunities for urban residents to indulge their love of the country without travelling far. Mobile farms are popular spectacles at fêtes, markets and child care centres, and annual Royal Shows around Australia still feature the best of the country among the Ferris wheels, arcade games and show bags. Yet there are many people who want their country experience to be a little more hands-on. They want the satisfaction and thrill of getting back to nature in their own backyard. And for them, there is no better animal to become acquainted with than the humble chicken.

It is for these people that this book has been written. For those who want to raise an animal in the confined space of their backyard. For those who want to teach their children how to respect animals and how fruitful this respect and engagement can be. For those who just want to get a bit dirty within easy access of a hot shower.

There are many reasons why chickens are the perfect animal to raise in your backyard. Let's have a quick look at some of these reasons. You might find several that are applicable to your situation, or you may have just one, overriding reason. Whatever your individual circumstances, the fact that you have bought this book means that you must at least be considering keeping chickens. So let's work out why.

FOR THOSE ON A BUDGET

One reason you don't see many cows in suburban backyards is that the cost of building a milking shed and purchasing all the elaborate milking equipment is beyond most hobbyists. There are other reasons of course but this is a book about keeping chickens, so we'll quickly move on to the subject of our attention. Chickens are reasonably cheap to keep. They don't take up much space (so you won't need to purchase the house next door and knock down the fence). Coops and runs can be built on a very tight budget. Feed is inexpensive, and so is the essential equipment needed to keep chickens content, healthy and laying – you can even end up turning a profit.

In short, chickens are the ideal choice for individuals, couples, families or share households that want the pleasure of rearing farm animals without exorbitant costs. Sounds good, doesn't it?

FRESH EGGS

The labelling on shop-bought eggs is becoming more and more complicated. As with most industries, marketing is paramount. That doesn't make things easy for consumers. Do you know exactly what conditions 'farm raised' hens live under? Are 'free range' hens fed the same hormone infused feed as caged hens, even though they get to stretch their legs? Will we soon get egg cartons labelled 'Reared in semi-detached, high-rise cages with scenic hillside views and classical music'? If so, would that necessarily mean that the eggs are any good?

The best way to ensure that you are eating fresh, healthy eggs from an environment that you approve of is to farm them yourself. And for that, you'll definitely need chickens. A cow just won't do the trick.

For a family that eats a lot of eggs, it can actually be economically beneficial to raise chickens for one's own supply of eggs. Even those who don't get through their daily supply should have no trouble selling their excess stock to neighbours and friends who have a semblance of concern about what they eat.

And the bottom line is that your eggs will taste a lot better, and the cakes you make will have a much richer colour and flavour, because your eggs will be fresher and will come from chickens that are fed what's good for them, not what's good for the corporation that owns them.

TEACHING RESPECT, RESPONSIBILITY AND BUSINESS SKILLS

Children born and raised in urban environments have minimal exposure to animals, except for domestic pets such as cats and dogs, and caged exhibits in zoos. Keeping chickens provides children with the opportunity to learn how to raise and treat animals with respect and responsibility. You do not need to be a professor of psychology to realise that children brought up to respect animals are more likely to respect them when they grow up; or that they will be far less fearful of animals than children who are not given such exposure. There is even some research that suggests a causal link between compassion for animals and care for fellow humans.

In addition, keeping chickens will teach your kids about business. At some time in their lives, just about everyone has to deal with the challenges, rewards, disappointments and elation that come with being a member of society. And one of the most daunting prospects is coping with the cut and thrust of the commercial world. Most children are taught basic skills, such as the benefits of saving pocket money, and the rewards that can be gained by putting in extra effort (such as washing dishes). But is this enough? Will the knowledge that 'a dollar saved is a lolly missed out on' lead to the rise and rise of Australia's next corporate colossus? Probably not!

Now take a short moment to think about the business lessons that can be learnt from keeping chickens. First, there's the crucial lesson that income generated must exceed expenditure. This is an easy lesson for children to learn. The money they receive from selling eggs must be greater than the cost of keeping the chickens. So they must price their eggs accordingly. Secondly, they need to market the eggs. This means identifying likely customers and finding ways to capture (and keep) these customers. And third, they can learn that reinvesting part of the profits can help grow the business and make it even more profitable. In other words, using some of the egg money to buy new hens will pay off big time. And you thought keeping chickens was just about getting back to nature. No, it's a lot more. It's about producing the next generation of entrepreneurs.

MAXIMISING THE OUTPUT

Making money from selling eggs is a great way of maximising the output of your chickens. But there is another way you can maximise the output. Chicken manure makes fantastic fertiliser for the garden. You can use the fertiliser yourself, to put more colour and zing into your roses, or you can bag it and sell it.

RIBBONS AND TROPHIES

Everyone loves a winner, particularly when that winner is *you*. And while you may have secretly rehearsed your prize winning Oscar speech, deep down you know the chances of standing on that podium accepting the trophy from Denzel Washington or Sharon Stone are not all that likely. But if you keep chickens, you can be in the running for a trophy or a ribbon from any number of shows around the country.

When people start keeping chickens, the thought of winning prizes is not generally the prime motive (or even an afterthought). But it's amazing how many people start taking the prize circuit more seriously once they have had a taste of competition. There are prizes for all types of chickens, so start rehearsing your speech. Just don't spend too much on your gown or suit.

PURELY AS PETS

And then, of course, there's the simplest reason of all to keep chickens: purely as pets. It is often said that there are two types of people on earth: dog people and cat people. What nonsense. There are plenty of people who find dogs far too much trouble to deal with *and* find cats too independent and wilful. There are families where one person is allergic to cat fur and another to dog hair. What other options are there? Goldfish owners spend more time visiting the local aquarium to replenish deceased stock than they do sitting in front of TV. Reptile owners get strange sideways glances from their neighbours. Rabbit and guinea pig owners are forever chasing their pets after they have escaped and then replacing the wire that has been chewed through. Owners of budgies spend so much time talking to their birds that they forget how to communicate with other humans. Pigeon fanciers are rarely invited to dinner parties. So what does that leave us with? It takes us back to the subject of this book: chickens.

And you'd be surprised what great pets chickens make. They are affectionate, more intelligent than you might think, and often hilarious. Adults and children alike will be constantly surprised and delighted by just how much joy chickens can bring.

LET'S GET STARTED

Can you believe there are so many good reasons for keeping chickens? Perhaps what's most astonishing, given all these reasons, is that not *everybody* keeps chickens. Which is just as well for you – you'd have no-one to sell your excess eggs to. Now that we've determined why you want to keep chickens, let's get started on the path toward having a backyard full of these intriguing, delightful, productive creatures.

20 Common Questions Answered

Whenever someone sets off on a new venture, they have plenty of questions. The only problem is, some of the questions can seem so simple that they are often embarrassed to ask them. Well, here are answers to 20 of the most common questions that people ask when they start keeping chickens. Many of these issues are discussed in more detail elsewhere in the book.

DO I NEED A ROOSTER FOR MY HENS TO PRODUCE EGGS?

The short answer is 'No!' Hens lay as many eggs without a rooster present as they do with a rooster present. You only need a rooster if you want the eggs fertilised in order to breed chicks. If you are not going to keep chickens for breeding purposes, you are almost certainly better off without a rooster for three main reasons.

First, hens often prefer an environment without roosters. While roosters offer protection from predators, a well-designed coop and run will make this unnecessary. Second, roosters can be hard to handle. They can be aggressive towards other chickens and their handlers. Ask yourself, do you really want to battle a belligerent rooster every time you feed your hens? The third reason is probably the most important: many local councils forbid the keeping of roosters in suburban backyards (see page 8, Are Chickens Noisy?).

DO CHICKENS SMELL AND ARE THEY DIRTY?

Do you start to smell when you haven't cleaned yourself for a while? Well, chickens are the same. If you keep the chickens' coop clean and make sure that they receive the food and care they need to remain healthy, then you won't have any problems at all. Some chickens may require special attention. For example, if you keep breeds that feature excessive feathering around their legs, you will need to make sure that dirt and mud does not accumulate. And breeds with particularly wide bodies often find it difficult to preen around the tail area. The solution is to know your breed and keep the chickens' coop clean. And before you ask the next, obvious question, 'How often should I clean the coop?', the answer is, 'No less than once a week'.

ARE CHICKENS NOISY?

Hens are not very noisy at all. They cackle a bit when they lay their eggs and sometimes when the sun comes up but they're certainly less disruptive than the barking dogs that every neighbourhood has. And if you really want to stop the hens cackling early in the morning, in order to remain on good terms with your neighbours, you can make sure the coop remains dark until a reasonable time. Now roosters are a different story altogether – which is why many local councils won't allow you to keep them. Roosters *are* noisy and there's no way to keep them quiet. And to make matters worse, they don't just crow at first light. They crow right throughout the day as well. If you are allowed to keep a rooster and you choose to do so, it may be a good idea to make sure that your neighbours receive a fresh supply of eggs each day. At least that way they'll think a little more fondly of you as they tuck into their scrambled eggs and toast at 5.30 a.m.

DO CHICKENS FLY?

Most chickens fly, though not very far. The ones that don't fly are the heavier breeds. You can safely keep them behind a 60 centimetre high fence in the knowledge that they won't abscond. The best fliers are bantams (small chickens) and the smaller standard breeds but even they are generally limited to flights of about 10–15 metres. Chickens are homebodies, so even those that do fly are only likely to fly onto a nearby tree branch to roost.

CAN I LET MY CHICKENS RUN AROUND THE BACKYARD?

Chickens often enjoy exercising and exploring the environment around their coop and run. Letting them out to roam around is a good idea. However, there are two main factors to be aware of. The first is the most important – predators. Chickens are not particularly good at fleeing predators. In fact they are downright hopeless. If there's no rooster to protect the rest of the flock, the chickens really only have you to look out for their wellbeing. That means making sure that the backyard is as secure as possible from dogs, cats and foxes roaming the streets. Apart from keeping predators out, you also want to make sure that the chickens stay in. Needless to say, don't let them out of their coop at night.

The second factor to consider is your garden. Chickens love scratching and pecking. You might appreciate your prize-winning rose bushes but to a chicken, the ground around the base of the bush is a veritable minefield of juicy insects just waiting to be uncovered. It's your call, but don't say you haven't been warned.

DO I NEED A PERMIT TO KEEP CHICKENS?

The responsibility for rules and regulations about keeping chickens lies with local government. Before going too far down the track of keeping chickens – in fact before you do *anything* about keeping

chickens – contact your local council. You will need to ask them the following questions:
- whether or not keeping chickens is allowed in your area
- whether or not you can keep a rooster
- how many chickens you are permitted to keep
- and whether or not you need the permission of your immediate neighbours.

ARE CHICKENS SMART OR DUMB?

It's all relative. They're dumber than humans but smarter than amoebas. To be honest, they're smarter than they're given credit for. They certainly respond to familiar routines and noises, particularly those to do with food. And they have a well-developed social ranking (see Is There Really a Pecking Order Among Chickens?, below) which is rarely deviated from. While you won't be able to discuss politics with your chickens, or teach them to play Tchaikovsky, you will be able to interact with them in a meaningful way.

IS THERE REALLY A PECKING ORDER AMONG CHICKENS?

The pecking order is a central part of life in a chicken coop. It determines how the chickens relate to each other. The pecking order is determined when a group of chickens first come together. They peck at each other, sometimes quite violently, until the order has been determined. Every chicken ends up with a position within the group, with the chicken at the top receiving the best of everything: food, roosting spot, shelter, etc. Any chicken that tries to usurp those above it is quickly put back in its place. When a new chicken is introduced into an existing group, it engages others in the pecking process until it has assumed its 'rightful' place. (Also see page 23, Pecking Order.)

WHEN DO HENS START TO LAY EGGS AND HOW LONG DO THEY LAY FOR?

Hens (or rather 'pullets' as young female chickens are called) usually start laying eggs when they are about five months of age. As for how long they lay for, that really depends on the breed and the individual hen. Generally, you can expect a steady supply of eggs from a hen for about three years, although the first year of laying is the most productive – it's pretty much downhill from then on. It's not uncommon for some hens to lay eggs for up to eight years, but the frequency with which they lay drops off markedly. Which brings us to the next question . . .

HOW OFTEN DO HENS LAY EGGS?

In general, chickens are bred for one of three reasons: for eggs, for meat or for show. However, there are a quite a number of breeds that are known as dual-purpose breeds. They are bred for both their eggs and their meat. The breeds that are bred just for eggs are, of course, the best layers. You can expect about three eggs every 4 days from them (about 270 eggs per year). Some of the dual-purpose breeds can deliver a similar number. The chickens bred for meat or show range from being fair layers (about 150 eggs per year) to poor (markedly less than 150 per year). You should also be aware that chickens will moult once a year after the summer and most do not lay eggs during this process. They also lay more eggs during summer than winter.

HOW OFTEN SHOULD I COLLECT EGGS?

Ideally, you should collect the eggs two or three times a day. Eggs start to deteriorate in heat and an egg collected and stored soon after being laid will remain fresher for far longer than one left for several hours. Also, eggs left for a long time are obviously more likely be accidentally cracked, either by a hen sitting on it or by one of the other chickens pecking at it or knocking it.

DO EGGS HAVE TO BE STORED IN THE FRIDGE?

No, eggs just have to be stored in a cool, dark, moist place. In fact, eggs will usually last longer in a kitchen cupboard than in the fridge because they will dry out earlier in the fridge. You can leave eggs for up to 5 weeks in a cool, dark, moist cupboard but only 2–3 weeks in the fridge. It's a good idea to write the date on which an egg was laid on the shell. That way you can eat them in order and not run the risk of leaving any too long. If you fear that an egg may have gone 'off', there is an easy way to test it. Just fill up a bowl with water and carefully place the egg in the water. If the egg sinks, it is fresh. If it floats, throw it out.

CAN I LET CHICKENS ROAM INSIDE MY HOUSE?

If you think there's something on TV that they'd like to watch, by all means invite them in. Just make sure you offer them a cold beer. Seriously, some people do allow their pet chickens to roam inside their house. They usually only have a couple of them and they restrict the chickens' access to the back of the house away from the plush carpets. But no matter how hard you try, you're never going to toilet-train your chickens.

DO CHICKENS MAKE GOOD PETS?

It all depends what you want out of a pet. If you want to take your pet for a walk on a leash, then no. But if you want a friendly, faithful animal to care for, to cuddle (yes, cuddle), to name, and to become one of the family, then you're on the right track. For many chicken owners, fresh eggs are a bonus. It's the animals themselves that they have fallen for. They're entertaining, companionable and cute. Of course, you can't dismiss the eggs. What other pets can provide you with something fresh to eat every day? Certainly not a dog, cat, guinea pig or goldfish. Plus, chickens stick to a routine, preen themselves and don't stray from their environment, so they are easy to look after.

HOW MANY CHICKENS SHOULD I KEEP?

The answer to this question is basically 'any number so long as it's more than one'. Chickens are social creatures, so they'll fret and be unhappy if they are alone. The final number you decide upon depends on many factors including the space you have available, the number of eggs you require (or can cope with), the amount of time you want to spend looking after your chickens, and, most importantly, the regulations enforced by your local council. When starting out, it's probably best to be conservative and start with just a few. That way you can get used to keeping a small number of chickens before filling up the coop.

WHAT DO CHICKENS EAT?

Commercial feed contains most of the nutrients that chickens require. But make sure that you give the right feed to the right chickens. Feed for hens contains calcium, to harden the egg shells, while feed for younger chickens does not. In addition to the commercial feed, chickens love leftover vegetable and fruit scraps. Most chickens also eat leftover pasta and rice, and they *love* cheese. In other words, most of the food scraps that you have to throw away will be consumed with great pleasure by your chickens. (For more information on what chickens eat, see page 77, Feeding and Watering.)

DO I HAVE TO BATHE MY CHICKENS?

No you don't. Chickens look after themselves from a hygiene perspective by taking dirt baths and preening themselves. The aim of a dirt bath is to get rid of mites, while the preening involves the chicken using its beak to clean its feathers of dirt and other unwanted particles. If you are going to exhibit a chicken in a show, you may wish to give it a wash and blow-dry to get it looking its shiniest, fluffiest best. The only problem you'll encounter is the chicken's natural dislike for being washed. Give yourself plenty of time and make sure you're at your patient best. In fact, when it comes to bathing chickens, four hands are better than two.

HOW LONG DO CHICKENS LIVE?

Treated well, a chicken should live for about 8–10 years. The main factors that will determine a chicken's longevity are the cleanliness of its environment, its diet, its treatment by other chickens, and its heredity. Its not uncommon for chickens to live a year or two beyond 10 years of age, and occasionally up to 15 years. Some chickens have lived up to 20 years but they are as rare as, dare I say, hen's teeth. Which brings us neatly to the next question . . .

DO CHICKENS HAVE TEETH?

No they don't, hence the common saying quoted in the question above. Instead of chewing their food as it enters their beak, they have a multi-stage digestive system that gradually breaks down their food.

WHERE CAN I BUY CHICKENS?

From poultry breeders big and small. Obviously, large professional outfits will be able to provide you with a wider range, a lot of advice and contact details for equipment and supplies. But it's also worth making contact with any neighbour who has hens and a rooster in case they want to sell you chicks that are hatched. Poultry clubs are the best starting point for a list of poultry breeders. See the resources section on page 120.

10 Steps to Keeping Chickens

There's no point building a chicken coop until you know how many chickens you want. And there's no point choosing a breed until you've found out if you're even allowed to keep chickens in your area. So here is a quick guide to venturing down the track of keeping chickens.

1. Contact your council and find out what the local laws are regarding keeping chickens.

2. As a courtesy, let your next-door neighbours know that you are thinking about keeping chickens. If they are concerned about noise and smells, assure them that chickens make very little noise (except roosters, of course) and that they are clean animals. And it doesn't hurt to offer your neighbours a supply of fresh eggs.

3. Measure the area of garden that you are going to allocate to the chickens. It has to house a coop and a run.

4. Work out how many chickens you can comfortably house within that space.

5. Determine the main reason you want to keep chickens (e.g. eggs, pets, show).

6. Research the breeds to find out which ones suit both your reason and the allocated space.

7. Decide which breed/s to keep.

8. Build or buy the coop and run.

9. Buy the chickens.

10. Enjoy them.

Know Your Chickens

A REWARDING EXPERIENCE

Having delved into the reasons why chickens make such superb backyard companions, the next step is to get to know chickens. This does not involve inviting a few chickens around for dinner and making small talk, but it does require some reading and analysis. The good news is that this won't take long, it's easy to understand and will make your experience with your chickens much more enjoyable, meaningful and rewarding.

ANATOMY AND PHYSIOLOGY

Many of the fowl-related terms in this book that you may be unfamiliar with relate to the anatomy and physiology of the bird. So let's start our 'get to know chickens' tour by taking a look at the inside and outside of hens and roosters, and becoming a bit more familiar with some of the anatomical and physiological features that you should know about (see Figure 1).

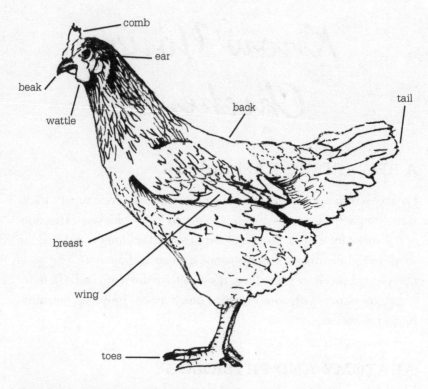

Figure 1 External features of a hen

The Beak

The beak is, as you might have guessed, used to collect food. It is made from protein and the top part of the beak can sometimes grow over the edge of the bottom part. When this occurs, some chicken breeders trim the top part of the beak in much the same way as one can trim fingernails. However, this can cause discomfort for the chicken at the time of trimming, and inconvenience later on if the result of the trimming inhibits the bird's ability to scoop food (see page 101 for details on beak trimming). Toward the back of the beak are two holes that are similar to human nostrils.

The Comb and Wattles

The comb and the wattles are the fleshy growths on top of the head (comb) and around the beak (wattles). Their essential purpose is to cool the chicken by re-directing the flow of air. The comb and wattles can also be used as a good indicator as to the general health and well-being of the bird. Loss of colour or shape in the comb or wattles is an indication that all is not well. Roosters have larger combs than hens.

There are seven types of combs: buttercup, cushion, pea, rose, single, strawberry and V-shaped combs (see Figure 2).

Figure 2 Three types of comb: single, pea and rose

The Ears

Chickens do not have external ears. They have small earlobes just behind and below their eyes. Finding the earlobes can be tricky as they are often covered by feathers. Amazingly, the colour of the earlobes gives an indication as to the colour of the eggs that a hen lays. Hens with white or very pale earlobes will lay white eggs; hens with brown earlobes will lay light brown eggs; while hens with black earlobes will lay dark brown eggs.

The Eyes

Chickens have very large eyes in proportion to the rest of their head, certainly larger than most animals. As a result, they have very good eyesight, far more acute than that of humans. This eyesight is particularly useful when chickens are scratching and scouring the ground for tiny particles of food. The eyes are a good indication of the health of a chicken – when healthy, the eyes are bright and beady, and without any discharge.

The Legs and Feet

Most breeds of chicken do not have feathers on the legs and feet, though there are some exceptions. Instead, chickens have scales to protect their legs and feet from the extensive digging and scratching that they do. Most chicken have four toes but there are a few breeds that have five. Male chickens have a spur on the back of the leg. The breeds that do have feathers on their feet should be kept away from muddy areas because the mud and dirt can become encased in the feathers.

The Feathers

Chickens have feathers for three main reasons. They provide insulation from hot and cold temperatures, and protection from harsh weather conditions such as rain. They also provide protection from injury. Finally, feathers provide chickens with the means to fly. However, most chickens are unable to fly further than a few metres at the most (see page 9, Do Chickens Fly?). They fly either to flee predators, to check out their immediate surroundings or to roost in a nearby tree. If you want to inhibit a bird's ability to fly, you can trim some of its feathers (see page 100 for details on wing clipping). This is common practice.

The Digestive System

A chicken's digestive system comprises nine main parts. These are: the beak, oesophagus, crop, proventriculus, gizzard, small intestine, large intestine, cloaca and vent. Several other organs, such as the pancreas and liver, also play vital roles (see Figure 3).

The digestive process begins once the chicken picks up food with its beak. The chicken has no teeth, so instead of chewing the food, it passes the food along the first part of the oesophagus to the crop. Here, the food is moistened and stored until the brain receives a signal indicating hunger. Once this signal is received, the food is sent to the proventriculus where enzymes and acid help break down the food further.

From the proventriculus, the food travels to the gizzard where muscles and grit combine to grind the food into a paste or mash. Once this grinding process is complete, the food travels through the small intestine where the protein and water in the food are absorbed. The remains of the food that are either indigestible or lacking nutrition then pass through the large intestine to the cloaca and vent, where they are emitted as waste. A final absorption process, mainly to capture any remaining water, takes place during the travel through the large intestine toward the cloaca and vent. The cloaca is the passage leading to the vent, while the vent is the orifice through which both waste and eggs are emitted. However, eggs take a different path to the vent than the waste does (see The Reproductive System, below).

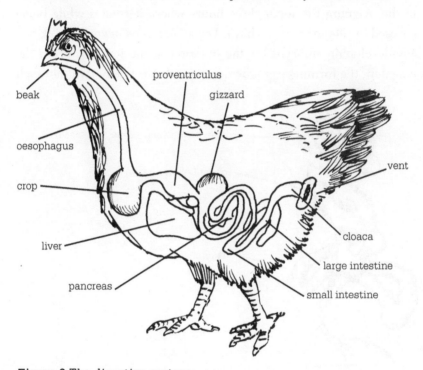

Figure 3 The digestive system

The Reproductive System

A hen's reproductive system comprises nine main parts. These are: ovary, oviduct, infundibulum, magnum, isthmus, uterus, vagina, cloaca and vent (see Figure 4).

It takes a hen about 25 hours to produce an egg. The process begins in the ovary where there are a cluster of follicles that resemble a bunch of grapes. When a follicle has reached its full size and maturity, it releases a yolk (which is the yellow part of the egg we are familiar with). The yolk drops into the entrance to the oviduct, which is a long, coiled tube that leads from below the ovary to the vagina. The egg is completely formed within the oviduct.

The infundibulum is located near the entrance to the oviduct and is where fertilisation takes place if spermatozoa are present. The yolk remains in the infundibulum for only about 15 minutes before travelling along the infundibulum to the magnum. The yolk remains in the magnum for about three hours where it rotates while being covered by albumen (egg white). The albumen protects the yolk and any developing embryos for the duration of the journey. From the magnum, the forming egg moves to the isthmus where two thin shell

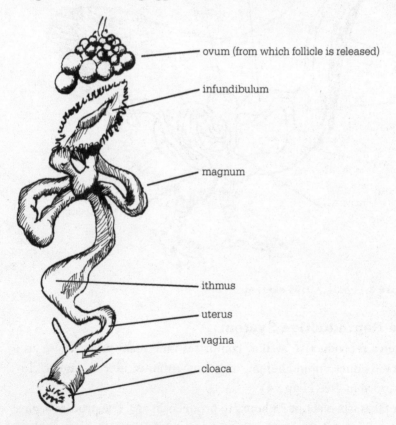

Figure 4 The reproductive system

membranes are positioned around the egg white. This process takes about 75 minutes.

The uterus is where the egg spends most time. It remains in the uterus for about 20 hours while the outer shell is formed around the membrane. Muscular activity forces the egg from the uterus to the vagina, where an anti-bacterial coating is added, and then to the cloaca and vent.

BEHAVIOUR AND PSYCHOLOGY

Although they have been domesticated for thousands of years, chickens have innate habits and behavioural patterns that date to their origins in the jungle. Some of these habits, such as a roosting instinct, aid the process of domestication, while others keep chicken breeders on their toes.

Pecking Order

Chickens are social creatures. In the wild, they travel in groups of about a dozen. Whether domesticated or in the wild, groups of chickens form a pecking order which provides them with their social position. A group of chickens put together for the first time may engage each other for days as they determine the pecking order. As the name suggests, much of this activity involves pecking one another. A rooster will always top the pecking order over a group of hens but the order within the hens is not so easily determined. While size plays an important role, it is not the sole determinant. While watching the process unfolding can be somewhat distressing to a human observer, it is best to leave the chickens alone unless one of them is in danger of being seriously harmed. Once the pecking order has been established, the group lives a fairly ordered life, interrupted only when new birds are introduced.

The pecking order establishes several rules, chief among them being that the chicken/s at the top of the order eat the most food. As a result they are often, though not always, the best layers. To ensure that your hens all receive enough food and water, you should provide several sources of each. Another mannerism common among hens at the top of the pecking order is to perch in the highest position. This is not necessarily a major problem but as these hens are also the last to settle down for the night they can unsettle other hens as they clamber to the topmost perch. Locating all perches on the same level alleviates this problem.

Aggressive Roosters
Roosters are hot-blooded males – they fight. However, the fighting abates after a pecking order among two or more roosters has been established. Occasionally, a young rooster will take on a rooster higher up the pecking order until the order is affirmed or changed. While such behaviour is normal, roosters do have spurs on the back of their feet with which they can inflict serious damage. If it appears that a rooster is being seriously hurt, it may have to be permanently removed.

Roosters use their aggressive nature and fighting quality to protect their group of hens and chicks. As a rule, they do not attack their human keepers but they will do so if they sense a threat. Their spurs can cause serious injuries, so roosters should be treated with a degree of caution at all times. Of course, roosters are not needed if you are keeping chickens for their eggs, rather than for breeding purposes. In fact, many local councils will not allow people to keep roosters in their gardens – mainly because of their loud crowing, rather than their aggression.

Cannibalism
Chickens have, by nature, cannibalistic tendencies. They will continually peck a wounded or vulnerable member of their group until they have killed and devoured the bird. Such behaviour can be the result of one of a number of factors including lack of nutritional feed, close confinement or boredom. Obviously, you should take steps to improve the conditions under which your chickens are kept if cannibalism occurs. This may involve giving the fowl more space to roam during the day, providing them with distractions such as toys, and checking the quality of the feed that you are providing.

Cannibalism can also occur when a small injury to a chicken causes blood to break through the surface of the skin. Any chicken with blood present should be removed from the flock until the wound has cleared up. In addition, any chicken that regularly appears to be engaging in cannibalistic activity should be permanently removed from the flock.

Moulting

New owners of chickens often get a shock the first autumn when their hens suddenly stop laying. If they look closely at their hens at this point, they will notice that their feathers have started falling out. This process is called moulting and it occurs every autumn, starting from the first autumn after they have reached maturity.

This condition is nothing to be concerned about, unless you suspect that the shedding feathers are a result of cannibalism rather than moulting. New feathers quickly replace the old ones. Moulting is a natural process that preserves a bird's chances of fleeing danger by replacing old and damaged feathers with new ones. Moulting hens may stop laying for up to three months. As a hobbyist, this should be of little concern to you. Some commercial operations use particular methods to get some hens to moult earlier than others in order to ensure the operation is always producing eggs. However, unless you need a continual supply of eggs, it is better to let nature take its course.

Broodiness

Broodiness refers to a hen's mood when she insists on doing nothing but sitting on a clutch of eggs. Fertilised eggs require about 21 days of incubation to produce chicks, and the warmth and protection from the hen are essential. Obviously, a brooding hen is no problem if you want to breed chicks. You just have to make sure that the hen moves from her nest at least once a day to eat, drink and defecate. If she doesn't, you will have to forcibly remove her for a period so that she does not starve or dehydrate.

However, a brooding hen can be a nuisance if you are quite happy with the number of chickens you have or have no rooster to produce fertilised eggs. In such cases, you have two options: remove the hen or remove the eggs. Removing the hen involves placing her in another pen where she cannot see or access her old nest. After a couple of days, she should have forgotten about the eggs. Just remember that a brooding hen will not want to be removed and is unlikely to be removed easily. Removing the eggs is another option but once you have done so, keep an eye on the hen to make sure she leaves her nest for food and water. Some hens are particularly broody and will remain on a nest even once the eggs have been removed.

Preening
One habit that often fascinates new chicken owners is preening. This ritual generally takes place in the morning. It involves a chicken using its beak to squeeze oil out of a gland near its tail and then spreading the oil on its feathers. This protects the feathers from the elements and helps rid the bird of mites.

Scratching
Chickens also spend an inordinate amount of time scratching the ground. They do so to look for food, such as discarded grain, worms and insects. You need to know this for one very important reason: do not plant your prize-winning flowers anywhere near your chickens.

Dust Baths
Just like many humans, chickens like to have a bath. But the big difference is, chickens bathe in dust. It would be nice to think that this is because they are extremely diligent at adhering to water restrictions but the reasons are somewhat less altruistic.

Chickens take a dust bath to clean themselves and to remove insects, particularly lice. They also do it to cool down. When they have a dust bath, chickens roll around, stretch out and generally enjoy the experience. If you don't have a suitable area of dirt accessible to your chickens, buy or make a sandbox and fill this with dirt.

The Right Chickens For Me

CATEGORISING CHICKENS

Chickens are either purebred or crossbred. Purebred chickens, as the term suggests, come from a long line of chickens that have not been mixed with any other breed. Crossbred chickens are a mixture of two or more breeds. Most purebred chickens are kept for showing, while crossbred chickens are used for egg production.

Chickens are categorised by class, breed and variety. The class of a chicken relates to the part of the world where the chicken originated, so, there are American, Asiatic, English and Mediterranean classes. Within each class fall a number of breeds. Chickens within a breed have similar features, such as shape, skin colour, feather position and number of toes. Breeds are divided into varieties according to features such as feather colour or comb type.

There is one other major classification that you need to be aware of. Chickens are also categorised as either standard or bantam. This relates to their size. Bantams are about one-quarter or one-fifth the size of standard chickens. Some bantams are small versions of standard fowl, while others are an exclusive breed.

The types of chicken that you keep will depend on various factors, including available space, the surrounding climate and environment, and your reason for keeping them.

CHICKEN BREEDS

Before we try to match the right chickens to your individual circumstances, let's take a look at some of the more popular breeds, along with some of the rarer but interesting breeds, from which you can choose.

(Refer to page 130 for a list of websites where you can view photographs of the various breeds).

Ancona
Originating in Italy, near the city of Ancona, this breed is characterised by its black feathers with white tips. Some have a single comb, others a rose comb. They are predominantly kept for laying eggs, being able to produce a large number of white eggs (220–240 per year). Two particular advantages of keeping Anconas is that they become very trusting after a period of gentle care and they are rarely prone to broodiness. They are, however, relatively good fliers and will need to be completely fenced in or have their wings clipped.

Andalusian
Originating in Spain, Andalusians are closely feathered birds that produce 200–220 white eggs per year and are not broody. However, they are primarily kept for show. The only problem with breeding Andalusians for show is that only blue feathered Andalusians can be shown and breeding a pair of blue feathered Andalusians always produces several white and black feathered chicks, as well as blue feathered ones.

Araucana
Originating in South America, most modern Araucanas have been bred with other breeds and have lost the distinctive ear tufts and floppy pea combs that distinguish the purebred version of this breed from most other chicken breeds. Araucanas are moderate layers (180–200 eggs per year) and produce eggs of various colours, particularly shades of blue and green. Araucanas are sometimes sold under the name Easter Egg Chickens.

Australorp
This chicken is an Australian bird developed from the English Orpington. It is well suited to Australian conditions and the female matures quickly, sometimes being able to start laying before five months of age. The Australorp has black feathers with a green sheen and lays more than

300 light brown eggs each year, making it one of the most efficient egg producers of all chickens. An Australorp hen actually holds the world record for laying the largest number of eggs in a year (364).

Barbu d'Anver
Also known as the Belgian Bantam, this chicken has no standard equivalent (unlike many bantams). It comes in various colours and is distinguished by its beard and extensive facial feathers. These facial feathers can impede its vision. Barbu d'Anvers are great pets for children, as they are small, easy to handle and have a calm demeanour (particularly the hens). They are also popular show birds, but are quite poor layers. Other closely related Belgian Bantams are the Barbu d'Everberg, Barbu d'Uccle and Barbu de Watermaal.

Barnevelder
Originating from Holland and named after the Dutch town of Barneveld, the Barnevelder has a reputation as a lazy bird and sometimes needs prompting to get the necessary amount of exercise. Mind you, that's no reason not to keep Barnevelders. We all have friends that fit that description (or we might even fit it ourselves!). While Barnevelders are only moderate layers (180–200 eggs per year), their eggs are a very rich brown colour, which makes it a popular breed. Their quiet disposition makes them good pets.

Brahma
Originating from Asia, Brahmas are extremely large birds. However, their size belies their nature: they are docile and make great pets. They lay brown eggs but are not among the most productive layers (150–180 eggs per year). Their abundant feathers make them particularly hardy in cold climates. Brahmas do not need to be kept behind high fences as they cannot fly. Brahmas are not among the most common chickens kept in Australia, though an increasing interest in this bird has led to their inclusion at most shows.

Campine

Originating from Belgium, Campines come in two varieties: golden and silver. Unusually, the hens and roosters both display the same colour pattern. Campine hens lay 160–180 white eggs per year, but are mainly kept for showing. They can be difficult to handle, being particularly flighty and active, so are not recommended for your first experience keeping chickens.

Cochin

Originating from China, the standard Cochins are among the largest breeds of chicken and have one of the most profuse plumages. They are reasonably docile and therefore make great pets. Their good nature and fluffy appearance made them very popular among British royalty in the 1800s. Their egg laying prowess, however, is less impressive (140–160 eggs per year). The Cochin's profuse plumage makes it a hardy bird in cold climates but it should be kept away from muddy ground as its plumage extends around the legs and feet, making it susceptible to developing mud balls. As with other large fowl, they do not fly and so can be kept behind a low fence. Unfortunately, their size, combined with their lazy nature, can cause heart problems.

Dorking

Originating from Italy but taken to England during the time of the Roman Empire, Dorkings have five toes on each foot and a large comb. They are reasonable layers (140–160 eggs per year) and are one of the few chickens to lay eggs of a different colour to their earlobes (see page 19, The Ears). Their earlobes are red, while their eggs are white, though sometimes they have a pink tinge. The five main varieties are White, Dark, Silver-Grey, Red and Cuckoo. Dorkings are not suited to cold weather but their gentle nature makes them great pets in suitable climates.

Faverolle

Originating from France, Faverolles are a very distinctive breed, with full beard, muffs, leg feathering and five toes, with one of the toes pointing up. As such, they are kept mainly for ornamental and exhibition purposes, though they are reasonable layers (160–180 eggs per year). Another attraction for those considering keeping Faverolles is their gentle nature. They can be very affectionate and are particularly suited to children.

Frizzle

Originating from southern Asia, Frizzles are among the two most distinctive breeds of chicken (see Transylvanian Naked Neck for the other). They have long feathers that curl backwards towards the head, giving them the appearance of having been rescued mid-cycle from a washing machine. Bantam Frizzles are more common than standard Frizzles and they are kept for ornamental and exhibition purposes. However, they are also reasonable layers (160–180 eggs per year). Because of this breed, the term 'frizzle' is also used to describe 'mutant' chickens that have feathers that curve towards the head.

Hamburg

Despite its German name, Hamburgs originate from Holland. They are small chickens but among the most productive layers, producing 200–220 small white eggs per year. They are active and very flighty, so need to have their wings clipped or be kept behind an enclosed fence. There are more than 10 varieties of Hamburgs, the most popular of which include the Silver Spangled, Golden Spangled, Gold Pencilled and Silver Pencilled.

Houdan

Originating from France, the Houdan is a large chicken that comes in two varieties: mottled (black and white) and white. This chicken certainly stands out from the crowd. It has a crest, beard and muffs, as well as five toes. Some varieties have a V-shaped comb, while others (such as those in Australia) have a butterfly comb. Their appearance makes Houdans a natural ornamental and exhibition bird but they are also reasonable egg producers (200–220 eggs per year).

Indian Game

Also known as the Cornish, this breed originated in Cornwall, England. It is a heavy bird with a broad, well-muscled body and short feathers that sit close to the body. It is not a particularly productive layer (160–180 eggs per year) but the hens do make good brooders. Indian Game are not recommended if you have a small area in which to keep your chickens, as they require a fair bit of space for exercise. They are also prone to mite infestation because their wide body and short legs make it hard for the bird to preen under its tail.

Isa Brown

This is one of the most readily available and popular varieties for backyard egg producers. The breed is a hybrid of the Rhode Island Red and Rhode Island White, and was initially developed in France in the 1970s ('ISA' stands for Institut de Sélection Animale). Isas are generally docile, friendly and good with children, and the best producers can lay 300 brown eggs in their first year.

Japanese Bantam

Japanese Bantams are pure bantams, with no standard equivalent. Originating in Japan, their most obvious feature is the tall, upright tail. This tail is especially pronounced in the male. The other feature that really makes them stand out is their short legs. They have the shortest legs of all chicken breeds. As such, they need to be kept in an area with short, dry grass. They can even be placed on carefully manicured lawns without doing any damage, because their short legs make digging difficult. Although the hens are good brooders, they do not lay many eggs. They are mainly kept for showing or as pets. Their even temper, small size and long lifespan make them ideal pets for children.

Jersey Giants

The name says it all. Jersey Giants are huge chickens – in fact, they're the largest of all. They were developed in New Jersey in the United States, in a bid to produce a large bird with plenty of meat. However, their slow rate of maturation has meant that they have never been commercially viable for that purpose. The hens produce 160–180 brown eggs each year but their size makes them very risky propositions as they can easily crack their eggs when brooding. There are two common varieties: Black and White, and Blue. The main appeal of the Jersey Giant is its size: it certainly presents an impressive sight.

Langshan

Langshans originated in China and appear to be extremely tall. This is due in part to their natural height but is accentuated by the fact that they hold themselves very upright, presenting a very stately appearance. They also have very long feathers. The most common variety is the Black Langshan, though White Langshans are also relatively popular. Langshans are extremely quick and active birds, and also good layers (200–220 eggs per year).

Leghorn

When it comes to popularity, no breed comes close to the Leghorn. And the most popular variety of Leghorn is the White Leghorn. Leghorns originated in Italy and their popularity stems mainly from their ability to lay a large number of white eggs and their almost complete lack of broodiness. Leghorns are very alert and active, continually pecking and scratching. If you are looking for a reliable, prolific producer of eggs, you cannot go past the Leghorn – they lay more than 300 eggs per year. However, their dislike of being handled can make them problematic as pets.

Minorca

Originating from Spain, then developed further in England, Minorcas produce very large white eggs and thrive on human contact. The most common variety is the Black Minorca but there are also Buff, White and Blue Minorcas. The Minorca's comb is larger than most other breeds and the bird comes in rose comb and single comb varieties. They lay 200–240 eggs per year.

New Hampshire

The New Hampshire originated in the New England area of the United States as part of a breeding programme involving Rhode Island Reds. A particularly broody bird, New Hampshire hens only lay 120–150 brown eggs each year. However, they are very easy to handle. The New Hampshire has a rich chestnut red colour and is commonly referred to as the New Hampshire Red.

Old English Game

The Old English Game originated in England, where it was bred as a fighting bird. It is probably the closest in looks and temperament to the Jungle Fowl, from which all breeds of chicken are descended. Today, Old English Game are kept mainly for show as their eggs are both small and relatively few in number (160–180 per year). Befitting their heritage, they have a natural fighting instinct, particularly among the cocks. It is recommended that mature cocks be kept apart. There are many different varieties of the Old English Game.

Orpington

Originating in England, the Orpington is a heavy set, amply feathered breed with a tendency to laziness. The hens are reasonable layers (200–220 eggs per year) and very broody. The main varieties are Buff, Black, White and Blue, with Buff being the most common. Their docile nature makes them good pets.

Pekin

This breed originated in China and is the bantam version of the Cochin. Like the Cochin, it is full-feathered. Pekin Bantams revel in human company and make great pets, but they're not fabulous layers (140–160 eggs per year). They come in many different varieties including Black, Blue, Buff, Columbian, Cuckoo, Lavender, Mottled, Wheaten and White.

Plymouth Rock

Originating in the US and named after the town of Plymouth, the Plymouth Rock is a popular farm chicken and makes an equally good backyard fowl. It is easy to handle and its hefty size means it cannot fly very high. The best known of this breed is the Barred, which features black and white barred feathers. The hens are reasonable layers, averaging 160–180 eggs per year.

Polish

The Polish, also known as the Poland, originated in Europe but its exact origins are unknown. Their fluffy, large crest gives them the appearance of having stepped out of the hair salon midway through a blow-dry. Some of the varieties are bearded. Their extraordinary look makes them popular ornamental birds but they're not particularly good layers (120–140 eggs per year). In cold weather, ice can form in their crests, so they need to be watched carefully at such times lest hypothermia set in.

Rhode Island Red

Originating in Rhode Island in the United States, the Rhode Island Red is one of the most popular breeds and produces 200–250 large brown eggs a year. It is very adaptable to a range of environments and, though large, is very active. As its name suggests, it has a deep red colour. Be aware that the roosters have a reputation for being overly aggressive.

Sebright

The Sebright is a pure bantam that originated in Britain. There are two varieties, Gold and Silver, and both varieties have rose combs. They are among the poorest layers (140–160 eggs per year) and are kept for ornamental and show purposes. Sebrights are extremely flighty and active, and not easy to handle. As such, despite their very elegant appearance, they are not recommended for beginners.

Silkie

Originating in Asia, probably China or Japan, the Silkie has 'cuddle me' written all over it. It possesses soft, fluffy feathers and has a gentle nature that makes it a popular pet for children. Smaller than a standard but larger than a bantam, the one problem with its feathers is that they are not waterproof. Therefore, the bird needs to be kept indoors during heavy rain. Silkie hens are not good layers but they are very broody. When selecting your first chickens, remember that the Silkie is a bit like the cute Labrador puppy that stares at you lovingly from the pet shop window – it is very hard to resist.

Sussex
Originating in Sussex, England, the Sussex is one of the more popular breeds. It is a good layer, producing 260–280 light brown eggs each year and is easily managed. It can adapt to various environments and is just as happy in a small space as free range. There are several varieties of Sussex, including Brown, Buff, Silver, Speckled and White.

Transylvanian Naked Neck
Also known as Turkens and just plain Naked Necks, these extraordinary looking chickens originated in Hungary but were developed further in Germany. They have about half the number of feathers as other breeds and are bare from the base of the skull to the shoulders, hence the name. When exposed to the sun, their neck turns red, like turkeys, but do not despair, this is not sunburn. Their reduced feathering makes them ideal in warm climates but they need extra care in colder areas. They are moderately good layers of brown eggs and have a calm and placid nature. One feature to be aware of is that the Naked Neck gene is dominant and crossing a Naked Neck with another breed will always produce a bird with a naked neck in the first generation. Another thing to be aware of is that choosing a Transylvanian Naked Neck probably says a lot about you.

Welsummer
Originating in Holland, the Welsummer is renowned for its large dark brown eggs, of which it lays up to 240 each year. The standards are large, active birds but not difficult to handle. The most common variety is the Black-Red Welsummer, though there are also Silver and Gold varieties. If your intention is to proudly show off the eggs that you are collecting, then you cannot go past the Welsummer.

Wyandotte
Originating in New York State in the United States, Wyandottes have 'curvy', fully feathered bodies. They are reasonable layers of brown eggs (180–200 eggs per year) and their docile nature makes them a popular backyard breed. There are many different varieties, including Black, Buff, Columbian, Gold Laced, Silver Laced, Silver Pencilled and White.

SELECTING YOUR CHICKENS

Now that you have read the main characteristics and idiosyncrasies of some of the more popular breeds of chicken, it's time to look at your particular circumstances and match a breed (or several breeds) to your purpose and environment.

Council Regulations
Before delving too deeply into the number and types of chickens that you should purchase, there is one very important aspect that you need to be aware of. While few councils ban chickens altogether, most have regulations on either the number of chickens you can keep and/or the distance they must be kept from your own and surrounding properties. You must also find out if you need to seek planning approval for the run and coop you are proposing to erect. The response that you get from your council may have an impact on the number or type of chickens you end up keeping.

The Right Number of Chickens
The number of chickens you keep will depend on three main factors: the amount of space you have for them; the amount of time you have to give to them; and the number of eggs you want.

In regard to space, allocate 1 square metre of floor space for each chicken (this will include space in both the house and the run). The more space they have, the better – up to a point. If there is too much empty space, the coop will be colder because the natural heat generated by the chickens has to fill a greater area. However, if your coop is too small, the mental and physical health of the chickens will suffer, leading to pecking, cannibalism and other unsavoury behaviour. If you only have a small area in which to build your coop, then you should strongly consider keeping bantams rather than standards. At about one-quarter the size of standards, bantams obviously don't need as much space per bird. They can make do with about 0.5 square metres per bird. For more information on how much space to allocate to your chickens, see Size and Space on page 46.

Keeping chickens requires a daily commitment. The more chickens you have, the longer you are going to have to spend looking after them. For a start, you should collect eggs two or three times a day. Collecting eggs from three hens is a lot quicker than collecting eggs from a dozen hens. And monitoring the behaviour of three or four chickens is a lot easier than monitoring the behaviour of three times that number. When starting out, it's best to take the conservative approach. If your ultimate aim is to keep 15–20 chickens, start with one-third that number until you have an effective system in place and are comfortable with the daily tasks that are required. Just be aware that introducing new chickens to an existing group will kick-start the process involved in establishing a new pecking order.

If you are keeping chickens primarily for their eggs, you can generally count on getting about three eggs every 4 days from good layers. If the eggs are just for your family, there's no need to have 15 hens, unless you intend eating omelettes for breakfast, lunch and dinner. Even if you are going to give some to your neighbours, they will probably not expect (or need) a basket of eggs every day.

Whatever number you finally decide upon, make sure that it's more then one. Chickens are social creatures and do not like being alone.

The Right Chickens for Your Environment

You've taken a look at photos of different chicken breeds and one stands out. That's great. But is it right for your environment? Some breeds cope well with hot weather, others are better in cooler climates. You need to gain a full understanding of how your preferred breed will cope with your climate and environment before purchasing any chickens. And the best people to ask are the breeders.

In general, bantams are your best option in hot climates because they are smaller and have less body mass to cool. Standards, being much larger than bantams, have more weight to move around in hot weather. If you are going to keep standards in a warm climate, stay away from the larger breeds such as Brahmas, Cochins, Houdans, Jersey Giants, Rhode Island Reds and Welsummers. Breeds with an excessive plumage are also unsuitable in hot climates, even bantams. The poor Cochin strikes out on this count as well. In cold climates, though, the large birds are very well suited.

It's not just the temperature that you have to be aware of. If you live in an area of extreme rainfall and your chickens will be forced to wade through muddy ground, stay well away from chickens with short and/or feathered legs.

Obviously, the best breed for many areas within Australia is the Australorp, a chicken bred for Australian conditions.

The Right Chickens for Laying Eggs

If your prime reason for keeping chickens is their eggs, there are breeds that are far more productive in this area than others. They include the Australorp, Leghorn and New Hampshire. If you have a preference for white or brown eggs, the Leghorn lays white eggs while the Australorp and New Hampshire lay brown eggs.

One general guide to selecting chickens for their egg production is to avoid the very large breeds. This is because they have usually been bred for their meat. Even the dual-purpose breeds bred for meat and eggs, such as Rhode Island Reds, do not produce as many eggs as the Australorp, Leghorn and New Hampshire.

And while bantams do lay eggs, they are obviously not as large as those laid by standard hens. This is not a problem if you are happy with smaller eggs.

The Right Chickens for Showing

Exhibiting chickens in shows requires an extensive knowledge of the characteristics that judges look for. Chickens being shown need to meet strict physical guidelines regarding everything from weight to feather colour. Chickens that do not meet these requirements are not eligible. In Australia, these particular characteristics are known as the Australian Poultry Standards and are published in a book available from state poultry associations (see the list of associations on page 121).

Not every exhibition and show will have a category for every breed covered in the Australian Poultry Standards, so if you are thinking of keeping chickens to show, make sure they are one of the breeds that regularly appear on the show circuit. The following list contains the categories for the Royal Sydney Show and offers a guide to other capital city shows:

- Ancona
- Andalusian
- Araucana
- Australorp
- Barnevelder
- Belgian, Barbu D'anvers
- Belgian, Barbu D'uccle
- Belgian, Barbu De Watermael
- Brahma
- Campine
- Faverolle
- Frizzle
- Hamburg
- Indian Game
- Langshan
- Leghorn
- Minorca
- Modern Game
- New Hampshire
- Old English Game
- Orpington
- Pekin
- Plymouth Rock
- Polish
- Rhode Island
- Rose Comb
- Sebright
- Silkie
- Sussex
- Welsummer
- Wyandotte

The Right Chickens for Children

As pets for children, bantams make much more sense than standards. Standards are heavier and therefore more difficult for children to handle. Many bantams have been especially bred for showing and are more likely to be docile. You cannot go past the adorable Silky Bantam for children. Other breeds to consider are Brahmas, Dorkings, Faverolles, Orpingtons, Pekins and Polish.

Housing Your Chickens

The most time-consuming, labour-intensive task facing anyone when they decide to keep chickens is designing and constructing the housing and exercising areas required. And even if the fowl house is going to be bought 'off-the-shelf' rather than constructed from scratch, there are still many factors that have to be considered. Before looking at those factors and then examining all the available options and requirements, it is important to understand the terms used when discussing housing for chickens.

DEFINITIONS

The main terms to understand at this point are: coop, house, run, nesting box, and roost.
- **Coop** refers to the entire area that will house the chickens. It includes the house and the run.
- The **house** is the structure in which the chickens will shelter, rest and lay eggs.
- The **run** is the outside area adjoining the house in which the chickens will exercise, scratch around and take dirt baths.
- The **nesting box** is inside the house and provides the hens with a cubicle in which they can lay their eggs.
- The **roost** is the structure on which the chickens perch at night. It is also inside the house.

BASIC NEEDS

When designing and constructing your chicken housing, keep in mind that there are several features that are prerequisites. Your house needs:
- a way for light to get in; and
- a way for air to get in.

The run and the house need:
- the means to keep chickens in;
- the means to keep predators out;
- the means to keep chickens safe from inclement weather; and
- to be comfortable enough for a human to enter in order to clean the area and look after the chickens.

There are two basic reasons why you need to house your chickens: to protect them from predators and to protect them from the elements. Keep this in mind when designing and constructing your coop.

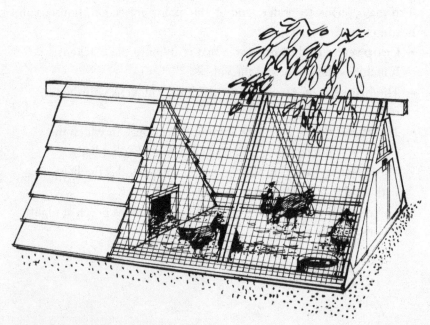

Figure 5 A simple chicken coop

Before going too far down this track, have a look at some coops in your neighbourhood. Most owners will be only too happy to show them off and to discuss the pros and cons of their coops, based on their experience. There is no point re-inventing the wheel. Learn from their mistakes and minimise future hassles (see Figure 5).

WHERE TO LOCATE THE COOP

The first step is to decide exactly where you want the coop to go. This is obviously a very important decision because once it's been built, it's hard to move (unless you have a mobile coop, which will be discussed shortly). You have to consider the elements, the lie of the land, your garden, your neighbours and your own proximity to the chickens.

If the land on which the coop is going to be built is sloped, you have to decide whether the top of the slope, the bottom of the slope or a level area is best. Positioning the coop on an elevated area of ground exposes the chickens to the worst of the natural elements such as rain and direct sun. On the other hand, positioning the coop in a depression leaves it open to running water and mud. Ideally, the coop will be located on a relatively level area of ground (see Figure 6).

As a considerate person, you will keep your neighbours in mind when working out where to locate your coop. In fact, your council may have specific regulations about positioning a coop, so ensure that you comply with these. You do not want the hassle of having to move or dismantle the coop just days after hammering the final nail. While hens do not make much noise, the further you position the coop from

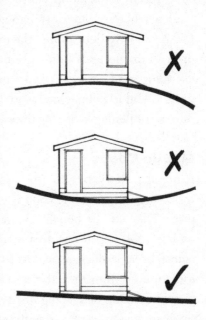

Figure 6 Positioning your coop

HOUSING YOUR CHICKENS   43

your neighbours, the less likely they are to complain about the noise that does occur. In addition, while they are clean birds, their scent will waft in the wind. Probably the most important reason to keep the chicken house as far away from the neighbours as possible is that chickens attract rats. Your neighbours may not be entirely comfortable watching rats scurrying across the middle of their pleasant game of croquet.

Having decided not to locate the coop on top of your neighbours, you then have to consider where to place it in relation to your own house. If the coop is right at the bottom of your garden, it's extra effort for you. And if you generally only visit the coop to collect eggs, you are less likely to notice any occasional problems that arise. If the coop is close to your house, then you will obviously notice any problems earlier and will be in a far better position to take preventive action. But do remember the aroma wafting in the wind. If the coop is too close to the house, your summer al fresco dining might not be quite what it used to be.

So ideally, your coop will be on a level area of ground, far from your neighbours but close enough to you to be observed regularly. Once you have determined the most suitable location and have come up with the necessary dimensions, lay some string on the ground (or use some chalk or spray paint) and make sure that everything is going to fit as you have imagined it to. Remember to take into account any outward-opening swinging doors.

Mobile Coops

A popular alternative to a permanent coop is a mobile coop. As the term suggests, a mobile coop is one that can be moved from place to place. They can be bought or made. The main advantage of having a mobile coop is that one area of your garden does not get completely ruined by your chickens. The main disadvantage is that it has to be physically moved from time to time. However, they are a great option for people thinking of keeping just two or three chickens, as they will be small enough to be moved with relative ease (see Figures 7a and 7b).

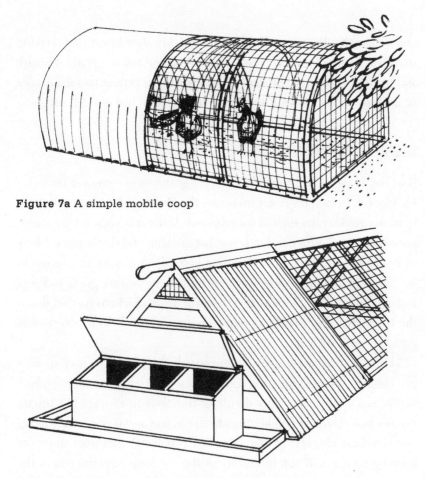

Figure 7a A simple mobile coop

Figure 7b Mobile coop with externally accessible nesting boxes

CLIMATE CONTROL

A coop in Antarctica will be very different to one located on the Equator. While your local climate is unlikely to be anywhere near those extremes, you still have to consider the climate when designing your coop. If you live in a moderate climate with little change from season to season, you'll probably be able to get away without having to insulate the house. But if your winters are particularly cool and/or your summers hot, then you may have to insulate against the cold and/or heat.

In Australia, it is preferable to position the coop so that the run and the entry to the house are facing south. This maximises the sunlight and means that the run will dry more quickly after heavy rain. Having maximum sunlight in the run is also beneficial for the general health and wellbeing of your chickens. They can always retreat to a shady area of the run or the shade of the house if they have had enough sun.

SIZE AND SPACE

If a coop is too small, your chickens are going to become very miserable indeed. In fact, it greatly increases the likelihood of them resorting to undesirable traits such as cannibalism. If the coop is too big, you've gone to a lot of effort and expense for nothing and the house is likely to be far too drafty. Before deciding how big you want your coop to be, you will have to decide how many chickens you are going to keep. Leave yourself some leeway to introduce more chickens further down the track, but don't build a coop to house 50 chickens if you're only going to start with six.

There are no hard and fast rules regarding how much space to allot per chicken, unless you are a commercial breeder using cages (which you're not). However, a good rule of thumb is to allocate 0.4 square metres per standard chicken inside the house and 0.6 square metres per standard chicken in the run. You can halve that area if you're keeping bantams. If you intend to let the chickens have the run of the garden on a regular basis, then you can provide them with a smaller run for them to exercise in between their romps in the garden.

One thing to remember at this stage (before it's too late) is that it is not only chickens that will use the coop. Humans have to get in there to collect eggs, clean and check on the chickens' welfare. Therefore, the house in particular must be big enough to enable humans to carry out these chores without having to continually stoop or crawl.

THE HOUSE

There's no place like home! While chickens don't necessarily have the same attachment to houses as humans do, they still have particular requirements and preferences. The first thing to say about the chicken house is that it has to be large enough to house the chickens comfortably (though not too large) and it must be tightly constructed to stop cold air blasting through gaps. It must also contain some essential equipment, such as a roost and a nesting box. Other than that, the sky's the limit, so to speak.

Some people start keeping chickens after their young children have grown up and no longer need constant attention and supervision. And sometimes such people have a children's cubby house in the back garden that is gathering dust and spider webs. If you are in this situation, you just might have the ideal chicken house staring you in the face. Is it tall enough for you to walk into? Is it large enough to hold the number of chickens you are considering keeping? Is it located on a level area of land? If you answered 'Yes' three times, then use the cubby house for your chickens. It will require some renovations but it will save you a great deal of time and money.

Disused (or underused) sheds are another possibility. If you have one, check that it's not too big and that the floor and walls are not rotting. If all seems okay in that regard and it's located in the right spot, then you too are on your way down the 'housing' track already.

Now, whether you have a cubby house or disused shed in your backyard or need to start from the very beginning, let's take a close look at all the features that you are going to have to construct or buy for your chicken house.

The Frame

Most chicken houses have a frame made from wood. That's because wood is relatively cheap, easy to use, versatile and easy to obtain. It also wears well when properly treated. However, when you paint or seal the frame (and any other parts of the house for that matter), only do so on the outside. Chickens are great peckers and won't be dissuaded by 'poisons' such as paint or sealant.

The frame of the house, and the components that fit into it, must be very tightly constructed. You want to reduce the amount of draft that can enter the house on wintry days. The need to install insulation may be alleviated if the house is constructed tightly enough.

One other tip to consider when constructing the frame is the benefits of having at least two separate compartments that can be both opened up and enclosed. There are likely to be times when one of your chickens is 'under siege' from one or more others and requires protection. It also gives you an area in which to isolate sick, old or injured chickens from the others.

The Floor

There are three types of floor that you can build your chicken house on: dirt, wood and concrete. A dirt floor is the cheapest and easiest, in that it's already there. Chickens love scratching and rolling in dirt and any grass that is present will soon disappear. All you have to do is build your walls around the dirt and you're halfway there. However, there are three main drawbacks to having a dirt floor. First, you are far more likely to have a problem with rats. Rats are burrowing creatures and a natural dirt floor will pose no problems. You might as well hang a sign on the door reading 'Rats Welcome'. A second drawback is that the chicken droppings and discarded scraps of food are harder to sweep up than on a wooden or concrete floor. And thirdly, a heavy rainfall can turn the dirt into mud if the house has not been located properly or if there is a drainage problem.

Wooden floors have the advantage of making it impossible for rats to burrow through the floor into the house. (However, that's not to say they won't find their way in another way.) Wooden floors are relatively cheap and easy to construct but again, don't treat them with paints or solvents on the side facing the chickens. The main drawback to wooden floors is that they will eventually rot, certainly before wooden walls will, because of the moisture from the ground.

Of the three options, concrete floors are the best for keeping out rats, the easiest to clean and the longest lasting. But concrete is also the most expensive and fiddly material to put down, and it's certainly the most permanent. If you have a dilapidated shed on an existing concrete slab, then by all means rebuild a house on the slab. But consider the implications of laying a new concrete slab in your garden very carefully indeed.

Whichever type of floor you eventually decide on, you will need to cover the floor with gravel (if it's a dirt floor) or absorbent materials such as chopped straw, wood shavings or hulls. This material is called 'litter'. Litter absorbs moisture and provides a comfortable resting place for the chickens. If you are only going to have a few chickens, or are housing them in a temporary location, you can strew the litter so that it covers the ground to a height of about 30–40 millimetres and replace this litter at least once a fortnight. However, the most effective system is the deep-litter system. The deep-litter system involves laying about 100–150 millimetres of litter on top of a layer of sand. Use a fork or rake to stir the litter every couple of days. You'll find that the height of the litter decreases because the chickens' activity and droppings cause the material to break down, so you will have to add more litter to keep it at the right height. Remove any piles of litter that have become damp and replace the entire amount once every few months. You can just throw it on the garden as fertiliser.

Doors
The first thing to say about doors is that there are two very different sized doors needed on a chicken house. One has to be the right size for chickens to go through, the other has to be big enough for people to enter and exit. You can get away with a smaller door for humans if you're only going to keep a couple of chickens in a small house, thereby requiring little cleaning. It can also be very useful to have a special door located near the area where you are going to dispose of the chicken waste. You can then sweep the waste straight off the floor and into the garden.

The door for the chickens should be about 30 centimetres wide and high. Obviously, it must lead straight onto the run. Make sure there is a very secure latching system attached. This is to protect the chickens from predators. If, as suggested earlier, you divide the house into different compartments, make sure each compartment has a door that can be latched. If the design of the house necessitates the door being positioned above the ground, leading either into or out of the house, then you may have to build a ramp for the chickens.

Some chicken owners have a stable-door arrangement, with separate upper and lower doors. Some use this system as a solution to having two doors – one for the chickens and one for the humans. Others use it because having an opening upper door gives them the opportunity to observe the inside of the house and the activity of the chickens without having to go inside and without running the risk of the chickens making a beeline for the door when it opens.

Windows and Ventilation

Windows and ventilation are required to light and air the house. It's not just for the good of the chickens but also for you. You need to see what you're cleaning without being overcome by fumes.

Windows can be opening and closing or fixed into the wall but having opening windows gives you more options when it comes to ventilation. However, make sure that the windows open vertically. Chickens will roost on just about anything and if the windows open out horizontally, the chickens will sit on the windows and you'll spend a great deal of time washing chicken droppings off the window panes (see Figure 8). It is also a good idea to cover the windows with wire or netting so that when they are open, predators cannot enter the house.

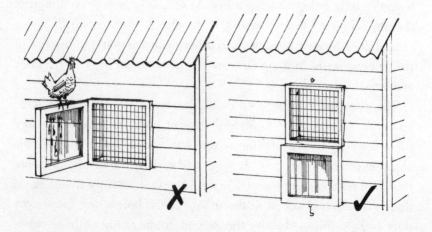

Figure 8 Opening windows – windows should open vertically

Ventilation slots or holes are important because of the chemicals, dust, gases and moisture that accumulate inside chicken houses. The health of your chickens will suffer greatly if they do not have proper ventilation. It is imperative that your chickens be kept cool in summer because chickens do not sweat and are not able to cool themselves easily. This is particularly important if you have a full-feathered breed. The ventilation holes or slots are best positioned across the top of the walls so that winter drafts do not stream directly onto the chickens. For the same reason, if you have a roosting system with high perches, don't have the roost directly opposite the ventilation holes. Ventilation holes should be covered in the same way as the windows, again to stop predators (and rats) getting inside.

The Roof

There are several roof shapes that you can consider for your chicken house, but the most popular are the lean-to (see Figure 9) and the double span (see Figure 10). The double span is most suitable for larger chicken houses, while the lean-to is great for smaller houses

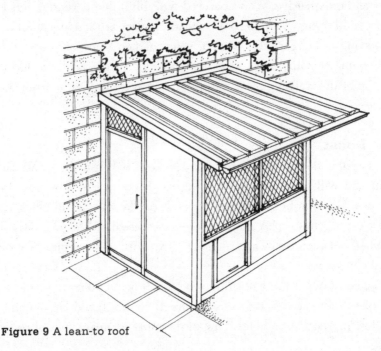

Figure 9 A lean-to roof

HOUSING YOUR CHICKENS

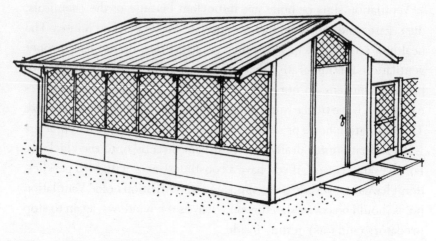

Figure 10 A double span roof

or for houses that are being placed next to an existing wall or fence. Remember to make sure that the roof has a fairly pronounced slope, to discourage chickens from sitting on (and fouling) the roof.

Aluminium and iron are suitable materials to construct a roof from, but can be expensive. Wood covered with bituminous roofing felt is also a good option. A tiled roof is obviously very attractive but you're not trying to increase the value of your chicken house as you would your home. Finally, make sure that the edge of the roof overhangs the top of the walls to allow rainwater to run off well away from the house.

The Roost

The roost is the place where your chickens will usually spend the night and will often rest during the day. It comprises a number of perches. This number obviously depends on the number of chickens you are going to keep. A single wooden bar raised off the ground and attached between two walls is enough if you are only keeping two or three chickens.

The roost needs to be positioned so that it is not going to be in the line of a draft – so don't position it directly opposite a door, opening window or ventilation holes. One of the most popular types of roost is a ladder (or two ladders joined together) placed against a wall (see Figure 11). Part of an old clothes horse will do the trick as well. You could consider a swinging roost as a way for the chickens to entertain themselves.

The required circumference of the perches will depend on whether you are keeping standards or bantams. Standards need about 4 centimetres of perch to curl their toes around, while bantams can do with 3 centimetres. The circumference of an average broom handle is perfect for bantams. Make sure that the perches have a rounded edge.

You should allocate about 30 centimetres of space per chicken per perch. This will help you determine how many perches you need on the roost. Allow about 50 centimetres between each perch and also between the wall and the nearest perch.

You also have to decide whether you want all the perches to be on the same level or to be positioned one above the other. It's obviously easier to have a couple of perches at the same height from the ground

Figure 11 Chickens roosting on a ladder

if you are only going to keep up to about six chickens. If you are going to have staggered perches, remember that the pecking order dictates that the chickens highest up the pecking order get the highest roosting positions. They are usually the last to settle down, so try to arrange the perches so that interruption is minimal.

Positioning the roost toward the back of the house will minimise disruption to chickens roosting during the day while others are using the run. If you are going to keep bantams, you can have the perches higher off the ground. Bantams love roosting up high. But if you have one of the larger breeds of standards, the lowest perch should be close to the ground.

Moveable roosts are a good idea because you are going to have to clean under them. If you attach the roost to a wall, it's helpful to be able to lift the roost and attach it to the ceiling when you want to clean underneath it. At the very least, make sure that you position the roost such that you are not going to do too much damage to your back by crouching to clean underneath it.

To aid the cleaning process, place a dropping pit under the roost. A dropping pit is a pan that collects the chicken droppings. It is covered in wire mesh and serves two purposes: it makes cleaning easier and it stops the chickens wading through their own waste. The holes in the mesh should be large enough for the droppings to pass through but small enough for the chickens to be able to walk on them without their feet falling through. The mesh cover should be easily removable for cleaning purposes. Just take the pan out, remove the mesh and deposit the waste in the garden. If the odour from the dropping pit becomes offensive in humid conditions, you can reduce the smell by sprinkling superphosphate or hydrated lime over the droppings.

Nesting Boxes

Another feature that you need in your chicken house is a set of nesting boxes. Nesting boxes provide your hens with a comfortable, clean, private environment in which to lay their eggs. In addition, they are a great aid to you. They save you foraging through piles of straw and wood shavings in search of eggs.

You do not need a separate box for every hen. Your hens will not become attached to particular boxes or spend too much time in the boxes (unless they are brooding). One box for every four or five hens is adequate.

The nesting boxes should be positioned about 60–100 centimetres off the ground, with a perch or landing platform for the hens to climb or land on. Consider placing them a bit closer to the ground if you have standard hens from one of the heavier breeds. The individual boxes should be about 30 centimetres high, wide and deep. In the bottom of each box place some straw, wood shavings or shredded paper to protect the eggs. Do not lay hay in the bottom of the boxes as it can develop a particular fungus that is detrimental to the health of your hens (see page 90, Aspergillosis). If you are going to put a roof over the boxes (and it's advisable, to keep the boxes nice and dark for the hens), make sure that the roofs have a steep slope to stop the hens roosting on them and making a mess that you could well do without. Also, test the structure to make sure that the box won't tip over if your hens walk or roost towards the edge. The larger standards can be very heavy and will not appreciate toppling a metre to the ground.

When your hens first see the nesting boxes, there's no guarantee that they'll immediately know what they're for. It won't take long for the penny to drop but if you want to prod them into action, place some fake eggs in the boxes. On the other hand, you don't want them to get too comfortable and start brooding. To avoid this problem, consider putting doors on each box and latching them at night.

As mentioned, collecting eggs from nesting boxes is easier than playing hide and seek with them among the litter on the floor. However, some people make it even easier by constructing the nesting boxes and the house with a flap/door that can be opened from the outside of the house. This way, there is no need to enter the house to collect the eggs (see Figure 12).

While the most popular system of nesting boxes is the cubicle arrangement, an alternative is the tunnel nest. The tunnel nest has no partitions (hens aren't that modest and don't really need privacy) and the hens can enter and exit from either end. One of the advantages of the tunnel nest is that it is easier to clean because there are no separate cubicles.

Figure 12 A nesting box that can be accessed from outside the house

Feeders and Waterers

There are many different types of commercial feeders and waterers available, and you can also make them yourself (see Figures 13 and 14). Decide on your feeders and waterers while you are designing the layout of the house, so that you can incorporate them. The first thing to keep in mind about feeders is that they should be suspended or attached about 15 centimetres off the floor. However, if you are going to have young chicks, you will need to put the feeders lower, or at least have one that the chicks can easily reach. If you are making your own feeders, make sure you include a lip to prevent spillage and wastage.

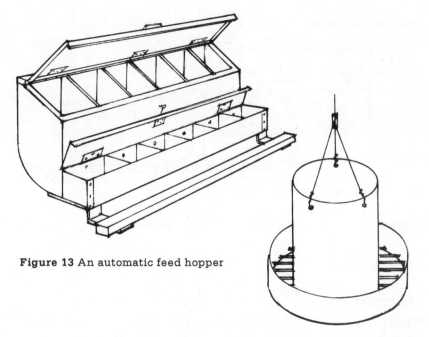

Figure 13 An automatic feed hopper

Figure 14 A hanging 'tube' feeder

A feeder that holds about 1.5 kilograms of feed will provide enough daily food for about 10 chickens. But it's wise to have two or three smaller feeders distributed about the house, so all the chickens can easily access food. You should have at least one feeder that can hold a week's worth of food for when you are away.

Suspended waterers should be suspended at the same height as the feeders. However, an alternative is a fountain system placed on the floor. The best thing about the fountain system is that large ones only need to be filled once a week, rather than daily like the smaller, suspended waterers. If you want to be particularly elaborate, you could consider a drinking nipple system where water only flows through the nipples when they pecked. It doesn't take chickens long to work out how to use it. The drinking nipple system (see Figure 15) is recommended by the Victorian Department of Primary Industries and the New South Wales Department of Primary Industries.

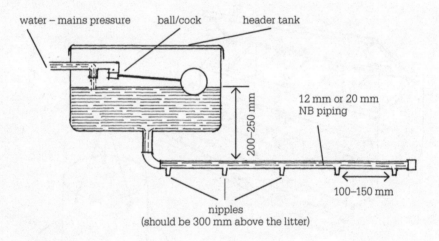

Figure 15 Drinking nipple water system

Distractions

Chickens are easily bored. That's why you need to give them plenty of space to roam, peck, scratch and generally please themselves. Chickens that get bored can start resorting to bad behaviour such as pecking and cannibalism. When it rains, they are usually restricted to their house, which can become cramped. Unlike humans, they don't have colouring books and Scrabble to entertain them on rainy days. However, there are alternatives that will keep them distracted enough to ward off boredom. These can include: broccoli stalks suspended upside down from the roof for the chickens to peck at; suspended mirrors that turn when pecked by a chicken; and a large pile of freshly mown grass for the chickens to explore.

Electricity

The final decision you are going to have to make before you erect your chicken house is whether or not to include an electricity supply. There's no doubt that electricity is useful, both for you and the chickens. Your decision may be based on your own work and lifestyle patterns. For example, if you work late and are likely to only collect the eggs once a day, after dark, a torch may not provide you with the light you need to find all the eggs.

Having artificial light in your chicken house has advantages other than enabling you to see in the dark. Hens' laying productivity is closely tied to the number of hours of sunlight. During winter, they lay far fewer eggs than in the spring and summer. However, the presence of artificial light in the house can increase the number of eggs being produced.

Electricity can also provide heating during cold winters. If you live in a climate that has harsh winters, then your chickens will greatly appreciate (and require) heating. You needn't install a heater, just a heat lamp or two. Apart from warming the chickens, the heat will also prevent the drinking water from freezing.

Once you've decided whether or not you are going to need electricity, you have two alternatives: running electricity under the ground and installing power points in the chicken house, or running extension cords directly from your house to the chicken house. Obviously, you can really only do the latter if the chicken house is located relatively near to your house.

House Designs

Now that you have some idea of what you need in your chicken house and what features have to be incorporated to keep the chickens safe, healthy and happy, here are a couple of designs that you could use as a starting point.

Perfect for Bantams

This design is perfect for people keeping a small flock of bantams (see Figure 16). It is an enlarged version of a doll's house. The materials and plans given here are to build one structure, which is all you might need if you only have a few birds (6–10). However, if you simply double the materials (except the dividing wall) you can make two structures (one the mirror of the other), join them together and have two separate breeding pens.

Materials

Note that all dimensions are in millimetres. Dimensions for framing timbers are nominal dimensions for Australian unseasoned hardwood.

3 ×	2400 × 1200 × 20 plywood (for walls and floor)
1 ×	2400 × 1200 × 12 plywood (for roof)
10 ×	50 × 50 strapping
2 ×	50 mm hinges (for door to yard)
2 ×	100 mm hinges (for door)
1 box	50 mm nails
2 ×	100 mm latches
1 ×	600 × 600 wire cloth (to cover window)
1 ×	100 mm door handle
12 ×	100 mm nails (for connecting walls to floor)
4 ×	cinder blocks (to raise floor off the ground)
12 ×	100 mm screws (for connecting cinder blocks to floor)
1 roll	bituminous roofing felt (e.g. malthoid)
3 ×	2400 mm metal flashing for edges of roof
1–2 tubes	caulking compound for seams along corners, roof and walls.

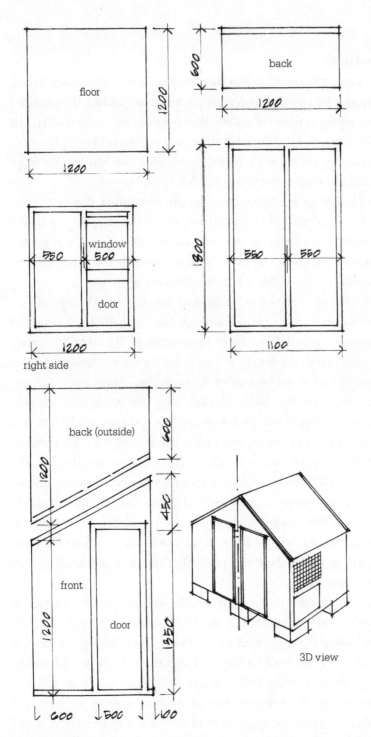

Figure 16 Plan

Instructions

Start by cutting the walls to size then cut the 50 × 50 strapping. Attach the strapping by laying it under the plywood and nailing the plywood to the strapping (instead of nailing the strapping to the plywood). The back wall needs to be spliced together, as it consists of two 'leftovers'. This is done by nailing it to the side walls once the building has been assembled and sealing the seams (as described below).

If two buildings are being constructed, remember that one is the mirror image of the other. Therefore, when the strapping is nailed to the walls it is placed on one side for the first house, but on the other side (outside) of the walls for the second house. This will put the strapping on the inside when the walls are set in place.

The width of the roof is 1200 mm but the length depends on whether there are one or two buildings. For one building, a length of 1500 mm is required. This allows for overhangs. If two buildings are being constructed, the length is 1400 mm for each. Metal flashing is placed on the sides and bottom of the roof/s in either case.

Set the floor on the cinder blocks, using one block for each corner of the building. If two buildings are being constructed, then two blocks are needed in the middle of the whole structure, plus one on each of the corners (six in total). Drive some screws through the floor and into the cinder blocks so that they are permanently affixed to the coop and serve as anchors. Make sure that the floor entirely covers the cinder blocks to eliminate the possibility of tripping on exposed blocks. Raising the floor on the cinder blocks allows for good air circulation under the floor, prevents dampness and helps reduce the problem of rats.

Once the walls and roof are cut out, the strapping has been attached to the walls, and the floor is in place, the building can be put together. Start by nailing the right wall to the floor. Once the right wall is in place, nail the front wall to the floor and then the right wall to the front. Do the same thing with the left wall. Once the front and side walls are up, nail the lower section of the back wall to the floor and side walls. Then put the upper part of the back wall in place and nail it to the side walls. Use the caulking compound to seal the seam. Also caulk all seams where the walls meet each other, the floor and the roof when in place. Staple the wire cloth over the window. Hinge both the door to the yard and the entry door, so that they swing outwards (attaching hinges to the 50 × 50 strapping, not the plywood). Place the roofing felt on the roof.

Once the house is up, it needs to be furnished. Attach your roost/s to either or both of the right and left walls. They should be about 750 mm from the ground and 600–900 mm long (for three bantams). Just inside the door, on the front wall, nail the nesting box/es. This location makes gathering the eggs easy. All you have to do is reach around and pick them from the nest. Like the roost/s, place the box/es about 750 mm from the floor. The feeder and waterer should fit neatly under the nesting box/es.

Vents should be cut into the front door and back wall. The vents should be approximately 100 × 300 mm. Locate one about 100 mm above the front door. Locate the other directly opposite on the back wall. The vents are high enough to prevent a draft on the birds when roosting but still allow for air exchange. You can also cut a window into the door. This allows you to look in at the birds without having to open the door. Both vents and the window in the door should be covered with wire cloth. Cover both windows with clear plastic during the winter; the vents are always left uncovered, for air exchange. Remove plastic in summer.

The real advantage of this design is that feeding, watering, and collecting the eggs can all be done from the outside. In addition, the house does not have to be entered for cleaning. All corners can be reached through the door. And because of the neat size and shape of the house, all birds can be seen at a glance.

Perfect for Standards

The following diagrams show the plan, front view and cross-section of a house suitable for about 12 laying hens or 24 pullets (see Figures 17a, 17b and 17c). It is recommended by the Victorian Department of Primary Industries and the New South Wales Department of Primary Industries.

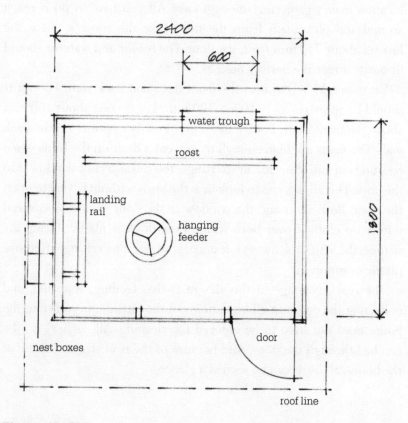

Figure 17a Plan

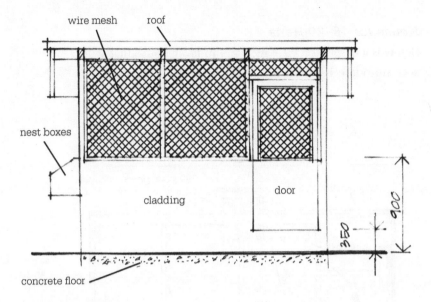

Figure 17b Front view

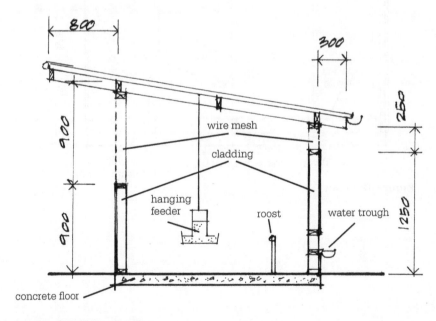

Figure 17c Cross-section

HOUSING YOUR CHICKENS

House for 15–20 Hens

Below is a design for a 2.4 m × 2.4 m layer house that will accommodate 15–20 hens (see Figures 18a, 18b, 18c).

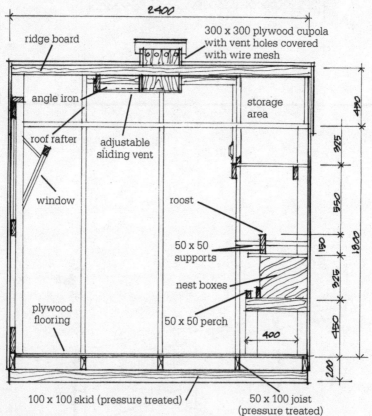

Figure 18a Plan

Figure 18b External view

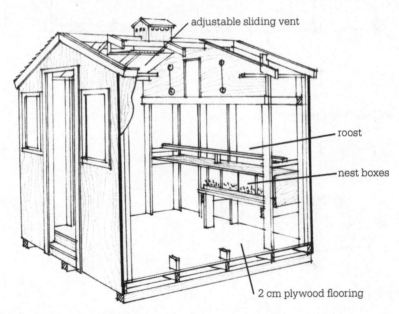

Figure 18c Internal view

HOUSING YOUR CHICKENS

House for 35–50 Hens

Below are plans for a 6 m × 6 m house that will accommodate 35–50 hens (see Figures 19a, 19b, 19c, 19d).

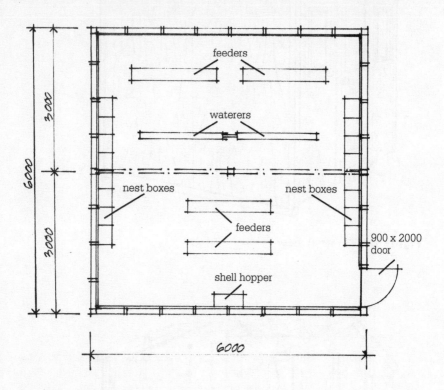

Figure 19a Plan

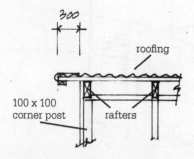

Figure 19b Eave section

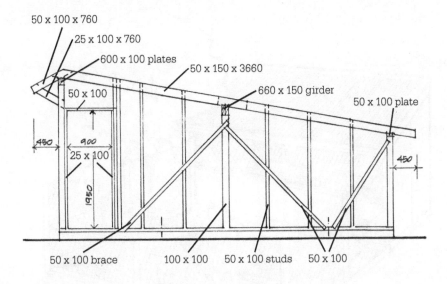

Figure 19c Side framing

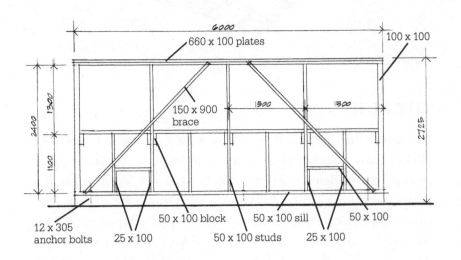

Figure 19d Front framing

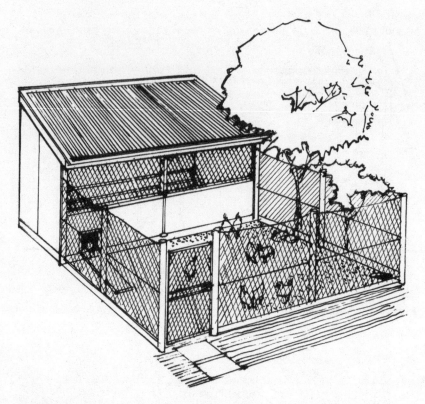

Figure 20 A large coop with run attached

THE RUN

You need a run for two reasons. You want to make sure that your chickens have a secure place in which to exercise and you want to protect your garden (see Figure 20). As far as size is concerned, 0.6 square metres per standard chicken is adequate (halve that for bantams), but you can make the run smaller if you're happy to let your chickens have the run of the garden sometimes. Whether or not they are going to use the garden, make sure there is always a generous amount of dirt in the run so the chickens can take their dirt baths.

The run should have an area that is permanently shaded, to give the chickens some respite from the heat on hot days. An innovative way of providing shade is to cut a hollow log in half, lengthwise, and place it facing down so that the chickens can walk inside it. Even the unshaded area of the run should have a wire or net roof to protect the chickens from predators and to keep them from flying out.

Another requirement to keep predators out is wire or netting under the ground right around the run. This can be achieved by extending the wire/netting that makes up the sides of the run so that it is buried about 10–15 centimetres below the ground or by adding extra wire/netting and attaching it to the sides. Make sure you bend it outward. Not only will this stop predators from being able to dig or burrow their way in but you'll also stop chickens escaping through holes they have created when scratching. And speaking of escaping chickens, even if you start with pullets and hens, you may eventually want to buy (or breed) chicks. Therefore, the holes in the wire/netting have to be small enough to ensure that inquisitive chicks don't explore their surroundings without supervision.

The other important factor to consider when designing your run is the benefit of having two runs. This helps to protect the ground where the runs are located. Runs get a lot of activity and the ground is scratched and pecked at continually. Switching the chickens from one run to the other will help it recover. The other option, if you only have a few chickens, is to build or buy a mobile coop that can be moved around.

Caring For Your Chickens

Buying your first chickens is a little like having your first child: it marks the beginning of a new era of responsibility. But while human babies grow up to be independent (some less or later than others), you are always going to be responsible for the welfare of your chickens. So whether it's feeding them or tending to them when they're sick, there's a lot you need to know. You will have a need to call on some of the information in this section regularly, and some of it rarely, but all of it adds to the store of knowledge that you will require to keep chickens. And what better place to start than the beginning of your contact with these beautiful fowl.

BUYING YOUR CHICKENS

You have four alternatives when buying your chickens (providing you're not after a rooster). These are: fertilised eggs, chicks, pullets and hens. Unless you have an incubator set up and want to throw yourself in at the deep end, stay away from fertilised eggs. You don't know what condition the chicks will be in, you don't know their sex, and hatching them is a lot of work.

Buying Chicks

Chicks, on the other hand, are a very viable option. They are usually sold as day-old chicks and as they have required little care, they can be picked up very cheaply (less than $5 each). You can purchase chicks as straight-run or sexed. Straight-run means that their sex has not been identified, while sexed means you know what you're getting (to about 95 per cent accuracy). While straight-run chicks are generally cheaper than sexed ones (because the breeder has not had to spend their time or pay someone to sex the chickens), if you are going to keep chickens for eggs, rather than to breed, always purchase female chicks.

The main advantage to buying chicks is that you get to see them grow up and become attached to them from the beginning. Chicks also adapt more easily to the new environment than older chickens and get used to being handled by you more quickly. The main disadvantages are the extended waiting period of several months before they start laying (and thereby saving or earning you money), the constant care they require and an increased mortality rate. However, don't try to anticipate a high mortality rate by buying more chicks than you're going to need – you could end up with too many chickens for your coop. Another way of ending up with more than you need is to succumb to the charms of day-old chicks. They're almost irresistible. Remember, they may be small now, but they'll grow and grow. Don't take more than you need.

Buying Pullets
Starting your flock with a small number of pullets is usually the best option. They may be more expensive than chicks but they are far easier to handle. In addition, the benefits of having your own eggs will become apparent very quickly indeed. You will probably have the choice of purchasing point-of-lay pullets or 6-week-old pullets. Point-of-lay pullets will cost a bit more but will start producing eggs within 2 weeks at the most. Six-week-old pullets have to be fed for several weeks before they become productive but they are basically independent. Whichever type you decide upon, keep an eye on them to make sure they do not succumb to diseases when placed in their new environment, and get confirmation at the point-of-sale that they have been vaccinated.

Buying Hens
There's little point in buying a hen if your objective is to maximise egg production. The first year of laying (the pullet year) is the most productive and it's all downhill from there. Having said that, there are bargains to be had if you are content with missing out on a hen's most productive period. Large-scale egg producers often sell a large

number of hens after their pullet year and will be happy to accept a small amount for them.

You can also miss out on the pullet year if your main objective is to show your chickens. By the time they have grown into hens, their features and characteristics are very distinctive indeed and you can see exactly what you're going to get.

Where to Buy Chickens

There are several options available to you. Start by talking to people in your neighbourhood who already have chickens. They may be able to save you a lot of time and direct you to your nearest breeder/seller. If you don't know of anyone nearby with chickens, you can contact breeding associations, feed stores or individual breeders. One way to identify local sellers is to look under 'Poultry Farmers and Dealers' in the Yellow Pages. Online and hard-copy classifieds are another option, particularly the *Trading Post*.

What to Look For

Before you buy your chickens, make sure that you are familiar with the local regulations regarding keeping chickens and that you have done your homework on the best breed/s to meet your needs, space, climate and lifestyle.

You should always carry out a physical inspection of the chickens before you buy them. While an expert will obviously be more likely to pick up undesirable traits, characteristics and conditions than a beginner, there are still many aspects that an untrained eye can notice.

First appearances count for a lot. Ask yourself the following questions: Is the chicken inquisitive or disinterested? Is it alert or listless? Does it look normal or does there appear to be a deformity? Do not purchase chickens that appear disinterested, are listless or have a deformity. Next, have a good look at particular parts of the chicken's body. Check its vent and tail area for signs of parasites. Make sure its eyes are bright and not watery. The feathers should be glossy, the legs strong and the breast firm. If you are able to tick all of those boxes, then the chances are your chicken will provide you with everything you want.

Preventative Health Management

When acquiring new birds, there are a couple of crucial points to remember that can make an enormous difference in preventing disease in your flock.

Common infectious diseases that an apparently well chicken may be carrying include coccidiosis (see page 92), Marek's disease (see page 96) and respiratory diseases, among others. Infectious parasitic worms are also common. Follow these guidelines to help prevent disease from entering your flock:

- If possible, buy vaccinated birds.
- Routinely treat new chickens for worms and coccidia before placing them in a clean coop, so they won't contaminate the pen by passing worm eggs and coccidia in their droppings.
- Always quarantine new chickens for at least 2 weeks before introducing them to your existing flock. In this way, if a disease becomes evident, at least it is only the new birds that you will need to treat.
- Ideally, avoid introducing juveniles into an established flock, as the older birds may carry disease-causing organisms (e.g. coccidia parasites) – the adults can have immunity to these organisms but the youngsters will not and disease could result.

Remember, it is generally the interaction of disease-causing agents, the environment and your chickens' immunity that determines whether a disease will occur – not just one of these things on its own. If you are new to owning chickens, it is worth taking your new birds to an avian vet for a health check to be sure that they are well and to assess whether the diet, husbandry and disease-prevention plan you have in place is appropriate.

If you ensure that you buy chickens from healthy stock, feed them well, provide a good environment and take sensible steps to avoid introducing disease-causing organisms, this will go a long way to ensuring your birds remain in excellent health.

FEEDING AND WATERING

You've got your chickens home and now you have to make sure they remain fit and healthy. That means providing them with a healthy, appropriate diet. And one of the first things to note is that chickens like to have variety in their feed. In that respect, they're no different to humans. Would you be happy eating the same thing day in, day out? Another crucial point to remember is 'what goes in comes out'. If you want your eggs to taste like those you buy in cartons at the supermarket, then limit your chickens' diet to commercially prepared feed. But if you want eggs that look and taste richer and fuller, then give them plenty of kitchen scraps and let them forage in the garden.

How Much Do They Eat?

The amount of food eaten by a chicken differs according to its size, breed, individual temperament and the time of year. As a general rule, a mature chicken will eat about 120 grams per day. However, you don't really have to measure and weigh the food. The best way to feed your chickens is by the free-choice method, which involves filling the feeders and letting the chickens help themselves. Unlike some animals (humans for example!), chickens will not overeat. They will graze throughout the day according to their needs. You don't have to feed them at particular times of the day, just top up the feeders when necessary.

Mash, Crumble and Pellets

Commercial feed provides chickens with all the minerals and vitamins that they require. It comes in three forms: mash, crumble and pellet. Other than the fact that chicks should be fed mash or crumble rather than pellets, because pellets are too big, it doesn't really matter which you go for. There is more wastage with mash because it gets slopped around more, but it is generally cheaper than crumble or pellets so there's no real cost difference. When you buy your feed, do so in bulk to save money. Buying it in small amounts will add a great deal to your annual feed bill. Store the bulk feed in airtight containers and it will stay fresh for weeks. However, do not keep it for longer than three months. If you have six hens, a 40 kilogram bag should last 6–8 weeks.

Mash tends to be the preferred option but you could consider having mash in one feeder and crumble or pellets in another. That way your chickens can have the best of both worlds. Your supplier will have different feed for chicks, young pullets and mature layers. The main difference between the feeds is that while pullets and hens need calcium to ensure that their eggshells are strong, calcium is detrimental to the health of chicks. You'll know if your layers need more calcium in their diet because the egg shells will be thin and fragile or virtually non-existent. Incidentally, an excellent source of calcium is ground oyster shell.

Some feed suppliers can provide organic feed, which comprises grain and other foodstuffs that have not had any chemicals used during growing. Such feed is more expensive than non-organic feed. If you want to give your chickens organic feed so that you can market your eggs as being 'organic', check the legal definition of organic with the food standards authority of your state government, as it differs from jurisdiction to jurisdiction and may even demand the use of unchlorinated water.

Grit
While it may sound a little distasteful to us, chickens need grit. This is because they don't have teeth. Grit is gravel or small rocks that help chickens break down their food. This process takes place in the gizzard. If your chickens have the run of your garden, they may well be able to forage enough grit naturally but if not, you should have a supply of commercially prepared grit available.

Healthy Extras
Your main feed costs will involve purchasing commercial feed. The healthy extras that you give your chickens will more or less be the scraps from your dinner table and garden. Most fresh greens, and vegetable and fruit scraps are suitable for your chickens, though there are some exceptions. Definite no-no's are citrus and potato peels,

tea leaves and ground coffee. Salty and sugary foods should also be avoided. Don't feed them foods that have a strong, lasting odour, such as garlic, because it will result in similar smelling eggs. Among the best types of greens and scraps are: lettuce, spinach, broccoli, tomatoes, corn, rice, pasta, bread, freshly mown grass and weeds.

The issue of feeding chickens meat is one that divides the chicken-rearing community. Most people do not feed meat to their chickens, and there is certainly a risk of diseases being passed on if you do so. However, some people have no qualms about including meat scraps in their chickens' diet.

Never feed your chickens food that has gone rotten as it could leave them very sick, particularly from botulism (see page 91).

One of the best ways to make sure your chickens are getting a varied, healthy diet is to let them roam around your garden. They will peck at the ground, eating insects, worms, seeds, grass, weeds and just about anything else that takes their fancy. Make sure you fence off the plants that you would prefer to remain untouched.

A special treat that has a similar effect on chickens as a piece of rich chocolate cake has on humans is scratch. Scratch can be bought commercially and is a mixture of cracked grains and corn. But like the chocolate cake, it is best in moderation. Too much and they'll get fat.

Water

Chickens (and their eggs) are about two-thirds water, so it's no surprise that they drink a fair bit of the stuff. In fact, on warm days, a chicken will drink about half a litre of water. That's a lot considering their size. Chickens don't just drink to replenish what they lose, they also need it to help them digest their food and to regulate their body temperature.

Chickens sip from their water supply throughout the day, so avoid having to continually top up the supply by making sure that you provide them with large waterers. You must make sure the waterers do not leak – not just to prevent wastage but because harmful bacteria quickly grows in puddles that have chicken droppings in them, which any puddles inevitably will. If a puddle develops, wipe it up immediately and find the source of the leak.

HYGIENE

A clean, dry coop won't smell. It also won't breed bacteria that attract flies and other pests. A clean, dry coop is essential for the wellbeing of your chickens. If you're going to commit to keeping chickens, it's up to you to make sure they have a hygienic environment.

Some cleaning tasks are daily tasks, while others are weekly, monthly or even annual. Let's start with the daily tasks. Every single day you are going to have to check the food and water supplies, get rid of food or water that has spilt into the trays or onto the floor, remove food scraps that have been left, mop up puddles, and briefly observe all your chickens to ensure that their appearance, activity and behaviour seems normal.

Once a week, you will remove all the manure from the run and the house, get rid of soiled litter, top up the litter on the floor, change the litter in the nesting boxes, and wash the insides and outsides of all the waterers and the feeder.

Once a month, you should thoroughly wash and disinfect the chicken house, using detergents and disinfectants bought from your supplier. Do not use conventional kitchen products. Once a month you should also check all of your chickens for mites, lice and other parasites.

Your annual job is to check the structure of your chicken house and the state of the fencing around the run. Is the wood on the house starting to rot? Do the floorboards have holes in them? Do the screws affixed to the roosts and the nesting boxes need tightening? Is the wire on the fence starting to unwind or come loose? At least it's only once a year . . .

HOT AND COLD

Chickens have to be watched carefully during extreme temperatures. Most feel the heat far more than the cold but precautions still have to be taken in cold weather.

The main problem with heat is that chickens do not have sweat glands and therefore do not perspire. But you can easily tell when a chicken is not coping with the heat because it pants with its beak open and holds its wings away from its body. You shouldn't let it get to that point though. On days when the temperature is going to be in the mid-twenties or higher, make sure that your chickens have an ample supply of water. The water should be in the shade so that it remains as cool as possible – you don't like drinking warm water on hot days and chickens are the same. In addition, you should fill a spray bottle with water and spray them at regular intervals.

Also make sure that the shading on the house and run is in place and that the ventilation holes are not blocked. Keep the windows and doors open to allow any breeze to flow through the house. Water restrictions permitting, hose or pour some water on the roof of the house and even consider turning a sprinkler on to provide some relief around the entire coop. If the temperature gets too extreme and your chickens are obviously suffering, consider placing a portable fan in the house.

Even when you have taken all possible steps to ensure your chickens will make it through the hot weather relatively unscathed, continue to keep a close eye on them. Make sure that none of them roost on the roof of the house, for the roosting bird can easily turn into the roasting bird.

The main precaution to take during the depths of winter is to make sure that the chicken house is not too draughty. The roosts and nesting boxes, in particular, should not be directly opposite ventilation holes, opening windows or doors. Artificial lighting in the form of heat lamps is recommended but make sure you don't place the globes close to straw or other flammable materials.

With cold temperatures can come heavy rain. Chickens don't like rain much and during winter may remain inside their house for extended periods. Consider rainproofing the run so that they can continue to exercise. Also, try to ensure that an area of dirt remains dry for them to take their dust baths.

CATCHING AND HANDLING YOUR CHICKENS

From time to time, for one reason or another, you are going to have to pick up your chickens. The best time to do this is at night when they are roosting. Even when they've noticed what's happening they'll be too drowsy to put up much of a fight. That's not to say you'll have a fight on your hands during the day. If you have had the birds since they were chicks or pullets and have developed a good relationship with them, handling them should not be too hard. Just approach them from behind and grab them on either side of their body, pressing their wings to their chest (see Figure 21).

If you have a recalcitrant chicken that you have to catch, try to force it into a corner or against a fence. Then pick it up by its legs or by the method described above. Don't lunge at it. Remain controlled at all times and even when running do so at a steady pace. If you have problems, just take a break and then try again. And be comforted in the knowledge that even the most experienced chicken handlers give up on a chicken from time to time. Some skilled handlers use a stick with a hook to catch chickens but this method is best left to the experts. Apart from the possibility of causing harm to a chicken, the sight of you in full flight with a pole and hook could be enough to send the rest of the flock into a panic.

Figure 21 The correct way to handle a chicken

EGGS

Now we come to the subject of eggs. For many people, this is the reason they have decided to keep chickens; they're sick of bland, commercial eggs and want eggs with more flavour and colour; they want to save money; or they just want the satisfaction of farming their own eggs. Maybe they're after all three. Whatever reasons you have for wanting to farm your own eggs, there's no doubt that it's a practice that's both satisfying for the stomach and for the soul.

What is an egg? An egg is approximately 70 per cent water, 10 per cent protein, 10 per cent fat and 10 per cent minerals (see Figure 22). It takes 25 hours from the time the yolk is released in the hen's ovary to the time the egg is laid. For a description of how each part of an egg is formed, see The Reproductive System on page 21.

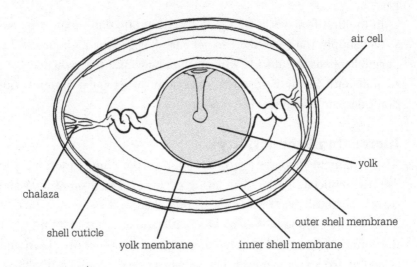

Figure 22 The internal structure of an egg

Preparing to Lay

Pullets start laying when they are about five months old. You may notice a pullet exhibiting some unusual behaviour prior to this occurring – it could be spending a fair bit of time hanging around the nesting boxes or constantly crouching, as it would if preparing for mating. If you do notice such behaviour, get ready for the first egg. But be warned, finding that first egg may not be easy. For a start, it's likely to be very small, perhaps as small as a grape, and probably not a perfect oval shape. And there's no saying where it will turn up. Some pullets dig a hole and lay in the hole, others lay in the grass, others lay . . . well, wherever takes their fancy. It's unlikely they'll lay their first egg in the nesting box. That takes a bit of time.

There are measures you can take to ensure that it doesn't take too long for your chickens to start using a nesting box. You can place a fake egg in a box and show it to them. They'll soon get the message. Fake eggs are available from poultry suppliers but a golf ball will do the trick. Another tactic is to lock the pullets inside the house until late afternoon, by which time they will have laid (if they're going to). This stops them getting into the habit of laying in the run or in your garden.

In the first few weeks of a pullet's laying life, don't expect a consistent supply. You're unlikely to get a fresh egg every 25 hours. If this period happens to coincide with autumn or winter than you may have to wait until spring before you can count on a regular amount. But don't despair, 'when it rains, it pours'.

Increasing Productivity

If you want to increase egg production in the winter months, you can use artificial lighting. The laying cycle revolves around light and there's obviously less of it in the winter than in the summer. Place a 25 or 40 watt globe in the chicken house and attach it to a timer. Set the timer so that the light switches on two hours before dawn and again for two hours at sunset. The only possible downside is that there will be more activity, and therefore more noise, at a time when you really don't want to hear it – and hopefully you won't.

Collecting Eggs

Ideally, you should collect eggs three times a day. Eggs start to lose their freshness as soon as they are laid. If you can't make it three times a day, the next best is twice a day, at lunchtime and late afternoon.

At the very least, collect the eggs once a day, in the afternoon. Apart from losing freshness over time, eggs are more likely to be cracked or even eaten by the chickens if they are left for too long.

Cleaning Eggs

When you first start collecting eggs, you'll notice that many of them have dirty shells. This can be a bit of a shock for people only used to buying cartons of eggs from the local supermarket. But think about it. While they sit there waiting to be collected, all manner of activity takes place around them. Dirt, dust and faeces seem to have a particular attraction to egg shells.

To clean your eggs, first wipe them with a dry cloth. That will get rid of dust and a light coating of dirt. But it will often not be enough to get rid of more attached materials. In these cases, scrub the shell with a coarse paper towel, steel wool or very fine sandpaper. Hopefully, this will remove whatever remains. Wherever possible, avoid washing eggshells with water. Eggshells have a water soluble coating that keeps bacteria from entering the egg. If this coating is washed away, the protection disappears and bacteria can run rampant. If you accidentally wash the outside of the egg, or do so as a last resort to rid the shell of firmly fused nasties, then eat the egg immediately. There'll be no harm done if you use it quickly.

Storing Eggs

Once you've collected and cleaned the eggs, you have to store them. They'll keep for up to about 3 weeks in the fridge and a couple of weeks longer in a dark cupboard. It's a good idea to write the collection date with a pencil on the shell of each egg. This makes it less likely that you will leave an egg for too long before eating it.

Be careful not to store eggs near foods that have a strong odour, or your eggs will give off a similar smell. During spring and summer, when your pullets and hens are in full flight and producing buckets of eggs every day, you may not get through the supply. If your neighbours also have more than they can handle, you can freeze the surplus stock.

(While on the subject of giving eggs to neighbours, you can use old egg cartons to store and carry them in. But you cannot use old egg cartons if you are going to transport eggs to sell. It is illegal for health reasons – the carton may be dirty and could therefore contaminate the eggs.)

Now back to the issue of freezing. To freeze eggs, break them into a cup or bowl and beat them with a fork. Add half a teaspoon of salt for every cup of eggs and then pour into plastic containers and put them in the freezer. Alternatively, pour into ice cube trays, freeze them and then transfer into plastic bags that can be tied or sealed, and place back in the freezer. Eggs will keep this way for 6–9 months.

One final point about eggs. Sometimes the yolk or white of an egg will have tiny blood spots in it. Don't worry about this. These spots are harmless and just the result of minor haemorrhages that have occurred while travelling along the hen's oviduct. Large-scale enterprises use technology to identify such blemishes before the eggs are packed and sold but it's purely for commercial reasons.

DISEASES AND HEALTH

It's inevitable that from time to time your chickens will exhibit symptoms of disease or display signs of erratic behaviour. It's important that you learn what to look for and what action to take. It's a responsibility that you should not take lightly – your chickens depend on you for their health and wellbeing. Even so, some of the diseases and conditions outlined in the following section can be fatal and you must be prepared for fatalities among your chickens, and even the prospect of having to cull affected birds for the good of the healthy ones.

While there are costs involved in seeking veterinary advice, doing so at an early stage can mean stopping an outbreak of disease and saving the life of one or more of your birds. Avian veterinarians routinely perform complex surgeries (e.g. for egg yolk related peritonitis) on pet chickens, but because of cost constraints it may not be appropriate to perform such operations on flocks. Vets experienced in dealing with birds can help to devise preventative medicine programs tailored to your particular circumstances. It is worthwhile establishing a relationship with an avian vet if you wish to give your birds the best possible care.

Vaccinations

Although vaccination will not always prevent disease in chickens, it can help to reduce occurrences. If you purchase your chickens from a commercial establishment, they should already have been vaccinated against Marek's disease as day-old chicks. Whether or not you need to provide additional vaccinations for older birds will depend on your local circumstances. Diseases against which vaccination may be recommended include fowl pox, infectious bronchitis, infectious laryngotracheitis and Newcastle disease. Unfortunately, vaccinating small backyard flocks can be problematic, as vaccines are typically sold in doses for 1000 birds.

If several of your chickens die at about the same time, or if you receive an alert about an outbreak of a particular disease in your area, it may be appropriate to vaccinate. Veterinary advice may be needed to diagnose what disease/s are present in your flock and what the best options are for treatment and prevention. If you do vaccinate your birds, always follow the instructions on the label. Use disposable syringes and needles and discard them after use, along with any unused vaccine, in a safe and secure manner. Do not use detergents or disinfectants near vaccination equipment and do not disinfect the chickens' skin before vaccinating with live virus vaccines (e.g. for fowl pox), because this may kill the vaccine virus.

Mites, Lice and Fleas

The most common external parasites of chickens are mites and lice, though fowl can also attract fleas. While lice will bite and cause major discomfort, they are not nearly as dangerous as the various types of mite that can, in severe cases, lead to the death of a chicken.

Mites appear as tiny red or brown dots, either crawling on a chicken's skin or clinging to the roost. The roost mites live on the roost and climb onto the chickens at night. There are also microscopic scaly-leg mites that appear as a crusty whitish scale on a chicken's legs.

Lice are easier to spot than mites because they are bigger. They are brownish-yellow in colour and have white eggs that stick to the feathers in clumps.

Dust baths are helpful for getting rid of mites and lice, which is why you should make sure there is always an area of dry dirt available to the chickens for that purpose. In addition, you should check your chickens every 2 weeks. Look around the vent and under the feathers near the wings and over the rump for signs of mites, lice and nits (e.g. eggs attached to the quills or barbs of the feathers). Also look closely at the roosts, particularly any cracks in the perches.

If you detect mites or lice, you have to take immediate action. The first thing to know about treating your chickens for mites and lice is that if one of your chickens has them, the chances are that most, if not all, of the others will be infected as well. So if you have to treat one, treat them all.

Purchase a spray or powder from your poultry goods supplier and apply it directly to your birds in accordance with the manufacturer's directions. It may be easier to do this at night when the chicken is roosting and more compliant. If mites have been detected on the roost, you can also brush or spray the application on the roost.

Scaly mites are not killed by topical sprays and powders used for feather mites, because they burrow under the skin. Because these mites need air to breathe, they can be killed by applying any non-toxic occlusive oil (e.g. paraffin oil) to non-feathered areas. Take care not to contaminate feathers and apply with caution – chickens have been killed by owners applying excessive amounts of oil or toxic anti-mite medications to treat scaly mites. It may also be necessary to also use oral or spot-on anti-mite drugs. Seek veterinary advice if the case is not responding to simple treatment.

There is one particular type of flea, the stickfast flea, which can cause more harm than other fleas. The stickfast flea attaches itself to a chicken, usually on the head or the breast, and doesn't move. It will suck the chicken's blood, eventually making it anaemic. As these fleas often attach to the eyelids, they can also cause blindness. Once detected, approved insecticide should be applied to the birds, and all litter in the house and the nesting boxes should be treated or destroyed.

Mites, lice and fleas are often transferred to chickens from other birds, so one way of minimising the chance of your chickens being infected is to make sure that they do not come into contact with other birds.

Worms

There are several different types of worms that can infect chickens. The most common are tapeworms, roundworms and hairworms, but the others include caecal worms, eyeworms, gapeworms and gizzard worms. The main difference between them is the area of the digestive tract that they inhabit, and whether their life cycle is direct (they're able to re-infect the same type of host they came from) or requires an intermediated host of another species.

Tapeworms attach themselves to the wall of the small intestine and can be up to 25 centimetres long. They require an intermediate host, usually a beetle or other insect, and are picked up when chickens eat insects infected with tapeworm eggs (which the intermediate host has obtained from contact with infected chicken droppings). Roundworms (*Ascaridia* spp) also live in the small intestine and are generally 5–10 centimetres long. They have a direct life cycle and are picked up when chickens ingest roundworm eggs present in the droppings of other birds. Pigeons can pass roundworms to chickens and vice versa. Hairworms (*Capillaria* spp) may have a direct life cycle or may be carried into coops by snails, beetles or other insects.

The main problem with having worms is that they lower a chicken's resistance to other diseases. Therefore, it is important that they be treated immediately and/or preventative action be taken. Routine de-worming treatment carried out twice a year will reduce the likelihood of worms appearing, though it's no guarantee.

Young chickens in a mixed-age flock may have poor immunity to parasites and become ill while other birds may not show any signs. It is best not to mix birds of different ages, to reduce the risk of worms and other diseases.

There are several symptoms of worms to look out for. These include poor laying, reduced appetite, increased thirst, diarrhoea, weight loss (for fully grown chickens) or not putting on weight (for young chickens). These symptoms can also occur with diseases such as coccidiosis or bacterial infections. Your avian vet can check your chickens' droppings to determine if worms are present, and can advise what medications are best to use and at what intervals. When diagnosed, approved preparations should be given to all chickens in your flock.

Apart from de-worming your chickens twice a year, there are other preventative steps that you should take. It is important that you do not let chicken droppings accumulate on the floor or litter and that your chickens be kept off damp areas. Dropping pits under roosts will help the situation (see page 54). If you suspect an outbreak of roundworm, keep the chickens out of your garden where they are more likely to find slugs, beetles and other possible sources of contamination.

Aspergillosis

Aspergillosis is a respiratory disease caused by a fungus in the genus *Aspergillus*. Known as brooder pneumonia, it affects young chickens and can be fatal. Symptoms include unsteadiness and nervousness. It must be treated immediately as it is highly contagious and can lead to the deaths of all young chickens in a flock fairly rapidly. If one of your young chickens dies after exhibiting the symptoms of Aspergillosis, contact your veterinarian and get treatment for your other young chickens. One precaution you should take against Aspergillosis is to make sure that you do not use hay as part of your litter system. Damp hay is a breeding ground for *Aspergillus* fungi.

Avian influenza

Avian influenza, also known as bird flu, is an infectious viral disease that has been known to affect humans. There are a number of types of avian influenza, some more virulent than others. While five outbreaks of avian influenza have occurred in Australia since 1976, none of these were of the strain that can be transferred to humans. The risk of a virulent strain occurring in Australia is relatively low. Nevertheless, authorities are on alert because such an outbreak could cripple the poultry industry.

Symptoms of avian influenza include: respiratory problems, swelling of the face, diarrhoea, decrease in egg laying, production of soft-shelled eggs, swollen combs and wattles, reddening of legs, and depression and listlessness. While these are common symptoms for many other poultry diseases, if avian influenza is suspected, contact your state department of agriculture immediately.

Botulism

Botulism is a form of food poisoning caused by ingesting a toxin produced by bacteria that may be found in rotting organic material. It is also known as Limber-neck because one of the symptoms is a bird's inability to raise its head. Other symptoms include diarrhoea, drowsiness, paralysis of wings and/or legs, and complete loss of appetite. The most obvious way to ensure that your chickens do not contract botulism is to make sure that their environment is clean and that their food and water are fresh and uncontaminated. Never feed chickens scraps of food that you believe are rotten. In fact, the best rule of thumb is to only give chickens scraps that have come straight from your table or that you would be willing to eat yourself.

If you suspect that one of your chickens may have botulism, provide it with fresh food and water (though it may not take any of it), and contact your vet to try to establish a diagnosis and the likely source of the toxin. There is no specific treatment for botulism, but some birds will improve with good nursing care and treatment for secondary problems. Be prepared for the worst, though; most chickens that contract botulism do not survive. However, if they are still alive after 48 hours, chances of recovery are better.

Bumblefoot

Bumblefoot is a fairly common infection that is characterised by the swelling of the pad on a chicken's foot. The swelling is usually accompanied by an abscess. To treat bumblefoot, wash the affected area, lance and drain the abscess, apply antibacterial ointment and bandage the foot. From an animal welfare perspective, it is best to seek veterinary assistance as soon as possible if your bird has bumblefoot, as cases caught early generally respond well to treatment, whereas poorly treated, neglected or advanced cases can be associated with severe joint damage and may not resolve.

Bumblefoot usually occurs when a chicken has caused damage to the pad of the foot by jumping onto or walking on a hard surface, allowing the introduction of nasty bacteria that triggers an infection. Vitamin A is needed for healthy skin and a deficiency can predispose chickens to bumblefoot – deficiency can occur in birds fed a seed- or grain-only diet. To minimise the chances of bumblefoot, make sure your chickens are not walking on hard surfaces, that the floor of the chicken house (particularly around the roosts and nesting boxes) always has a thick covering of litter, and that your chickens are getting a balanced diet. It's also worth checking that there are no splinters on the roost that might open a wound in the pad.

Chronic Respiratory Disease

Chronic Respiratory Disease (CRD) is a common respiratory disease among chickens and though highly contagious, it is rarely fatal. The symptoms of CRD include sniffing, sneezing, coughing and nasal discharge, with pullets most at risk of contracting the disease. If by some chance you don't notice the sniffing, sneezing, coughing and nasal discharge among your chickens, you'll certainly know that something's not right when you go to collect their eggs. Egg production can fall by as much as a third when pullets and hens are suffering from CRD. Don't wait for this disease to work its way through your flock (which it probably will do) because it is likely to recur in chickens that have already had it. You will need to treat your flock with antibiotics from your vet.

Coccidiosis

Coccidiosis is a disease caused by parasites that live in the gut wall of their host. It is very rare among chicks up to about 3 weeks old but is especially prevalent in chickens 1–4 months of age. Coccidiosis is usually contracted from the faeces of infected chickens and it can quickly decimate a flock of young birds. In fact, it is currently one of the costliest diseases within the Australian poultry industry. Symptoms of coccidiosis include bloodstained droppings, a floppy head and drooping wings, and ruffled feathers. Obviously, a clean chicken house with droppings regularly disposed of will reduce the chances of an outbreak. You can take other preventative action too. Live vaccines and feed containing coccidiostat can be given to young chickens to build up their immunity to the disease. But if there is an outbreak, the chicken house and all its fixtures, all equipment that you use, and even your own shoes and boots must be thoroughly cleaned before the chickens are returned to the house. Your vet will be able to test your chickens' droppings under the microscope to see if coccidia or worm eggs are present (both are tested with the same procedure) and can advise on the best options for treatment and prevention.

Crop Bound
Chickens become crop bound when long grass, string, litter or other materials form a ball and get stuck in the crop (see page 20, The Digestive System). It is noticeable as a large bulge in the bird's chest, particularly in the evening after a full day of feeding. Obviously, it requires urgent attention. The first thing to do is to try pouring a lubricant, such as vegetable oil, down the throat of the chicken in an attempt to get the object to move. At the same time, or if that doesn't work, you can massage the lump in a bid to break it up and loosen it. Hopefully, your actions will do the trick. If not, you should take the chicken to your vet.

Egg Bound
The best way to imagine the discomfort and anxiety that a hen feels when it is egg bound is to think of yourself as a pregnant woman unable to give birth. An egg-bound hen cannot pass eggs through her oviduct (see page 21) and, as a result, eggs may bank up much like peak-hour traffic. Without intervention, the hen will usually die. The best way

to avoid egg binding is to make sure that your hens have a healthy diet, including ample calcium (needed for contraction of the oviduct as well as formation of the shell), and that they get plenty of exercise. When hens become obese, a thick layer of fat develops along the oviduct, reducing the space for an egg to pass. However, egg binding can occur for a combination of other reasons, including when an egg is just abnormally large, if the hen is unwell or low in calcium, if it's the start of the laying season, or if the weather is cold. The most obvious sign of egg binding is if you see a hen making several unsuccessful attempts to lay an egg or moving in an obviously uncomfortable manner.

If you suspect egg binding, take immediate action. Warmth is sometimes enough to coax an egg that is not too firmly stuck. Place your hen under a heat lamp and massage the area around the egg (which you will be able to feel). Giving calcium liquid orally can help in some cases. If these methods don't do the trick, lubricate the vent with petroleum jelly (avoiding contamination of the feathers) and reach inside with a clean index finger as far as you can while massaging the hen's stomach. You may be able to coax or pull the egg free. If you are still unable to get the egg/s out, take the hen to your vet immediately.

Egg Yolk Related Peritonitis

Egg yolks are normally produced by the ovary and then picked up by the infundibulum (funnel end) of the oviduct. The egg white, membranes and shell are deposited as the egg passes down the oviduct before being laid. If this process does not occur correctly, egg yolks or partly formed eggs may end up in the abdominal cavity, where they can trigger an inflammatory reaction and cause what is know as egg yolk related peritonitis (EYRP).

EYRP usually occurs in older hens and can be associated with cancer in the oviduct or ovary, or it may be caused by scarring or inflammation in the oviduct. Hens with this problem will show a distended, generally fluid-filled, abdomen. Once a hen shows signs of EYRP, it is unusual for them to ever lay again, so if eggs are why you keep your chickens, euthanasia may be the path to follow. If your hen is a family pet and egg laying is irrelevant, then you should take the bird to your avian vet. Some hens will respond to draining of the abdomen and other medical treatment, but more often major exploratory surgery is required. In some cases, inoperable cancer will be found to be the cause, while in others the problem may be able to be resolved.

Fowl Pox

Fowl pox is a viral disease for which there is no treatment. However, vaccines are available and can be used to minimise the chances of fowl pox infecting your flock. If you are going to vaccinate against fowl pox, vaccinate all your chickens at the same time, as the vaccine contains a small amount of the virus. Fowl pox is spread by mosquitoes, so keeping your chickens' environment as mosquito-free as possible will also help.

The most obvious sign of fowl pox is the presence of wart-like lesions or scabs on a chicken's head, legs or mouth. They also accumulate in the throat and can severely disrupt feeding and breathing. Sometimes a chicken with fowl pox will also have a slightly swollen or runny nose. Some people recommend culling infected fowl to prevent it spreading, but as the lesions/scabs appear several days after the original infection, this action may be of limited use.

Frozen Combs

In particularly cold conditions, some breeds of chicken may experience frozen combs. As the name suggests, it is a condition that involves chickens' combs and wattles freezing in sub-zero temperatures. If you live in areas that experience such a climate, make sure that you eliminate drafts and moisture in the chicken house during the depths of winter. And if your chickens experience frozen combs, gradually thaw the combs with cold water and by rubbing Vaseline on the affected area.

Infectious Bronchitis

Infectious bronchitis (IB) is a highly contagious respiratory infection that causes coughing, sneezing, nasal discharge and breathing difficulties. If contracted by young birds, it can affect the reproductive organs. In layers, it can seriously affect egg production. There is no treatment for IB but a vaccine is available. If you suspect IB in your flock, you should seek veterinary advice to try to establish a diagnosis and to decide whether vaccination is an appropriate option.

Infectious Coryza

Infectious coryza is an acute respiratory disease not unlike the common cold in humans. And like in humans, it is most likely to appear in individuals that are run-down or tired, and among groups where overcrowding occurs. A healthy diet and environment are the best precaution. The main symptoms are nasal discharge, sneezing and swelling of the face under the eyes. If one or more of your chickens do contract infectious coryza, isolate them from the healthy birds, keep them warm (25°C), give them good nursing care, increase the amount of fresh greens they are eating, and give them Vitamin A boosts in the form of cod-liver oil or another approved treatment. If this does not resolve the problem, seek veterinary advice. Infectious coryza will generally improve with antibiotic treatment.

Infectious Laryngotracheitis

Infectious laryngotracheitis (ILT) is a respiratory disease characterised by coughing and sneezing in mild cases and gasping, neck extension, squinting of the eyes and conjunctivitis in more serious cases. In serious cases, death can occur by suffocation. There is no effective treatment for ILT but a vaccine is available. If you suspect ILT in your flock, you should seek veterinary advice to try to establish a diagnosis and to decide whether vaccination is an appropriate option.

Marek's Disease

Marek's disease is a viral disease that causes tumours in the nerves and some organs, including the ovary, liver, kidney and heart. It is most likely to affect chickens up to 6 months old. The main symptoms, which are hard to miss, are dropped wings and paralysed legs. There is no treatment for Marek's disease but chicks should be vaccinated at day-old stage. In fact, you should refuse to buy any chick unless you have proof of vaccination. If one of your chickens contracts Marek's disease, it should be destroyed, which will hopefully stop the disease from spreading. However, if several of your chickens contract it, it may be appropriate to cull your entire flock, then clean and disinfect the house, all equipment and fixtures, and purchase new, vaccinated, birds. Vaccination does not eliminate the chance of contracting Marek's disease, but it greatly reduces it.

Newcastle Disease

Newcastle disease is a highly contagious viral disease that can affect the digestive, respiratory and nervous systems. Chickens of all ages are susceptible and they can contract either a mild strain, which causes few symptoms, or a very serious one that brings on paralysis and even death. There is no treatment but a vaccination is available; in fact, it is compulsory in some states.

Prolapse

Prolapse is not a disease per se, rather a condition. It appears most often in overweight hens and involves the lower part of the hen's reproductive tract extending out of the vent during the process of laying. Usually, the tract retracts back inside after laying the egg but when this does not occur the condition is known as prolapse. If this occurs, you can attempt to push the distended organs back in. First, wash them with warm saline solution (1 teaspoon of salt to 1 litre of water), then smear a bit of Vaseline on them and push them back inside. Monitor the hen very carefully because a hen with prolapse is likely to be a target of cannibalism. If the individual bird is not valuable and the problem is not resolving, then the bird should be humanely culled. For valuable or pet chickens, veterinary attention should be sought early – some cases will respond to surgical treatment (various techniques are used).

Pullorum

Pullorum is a bacterial disease that affects chicks. It is caused by a bacterium named *Salmonella pullorum* and it passes from a hen to its chick through the egg or from one chick to another. The main symptoms are diarrhoea, droopiness, laboured breathing and ruffled feathers. Mortality among chicks with pullorum is high. Though careful care, observation and medication can sometimes save affected chicks, they are likely to remain carriers of the disease and can pass it on to other birds. Pullorum tests can be performed on fertilised eggs and are routine at large breeding establishments. If you are buying chicks, you should only do so from hatcheries that carry out such tests.

Tuberculosis

Tuberculosis is a slow spreading, but often deadly, bacterial infection characterised by gradual weight loss. It mainly affects older chickens and is rarely encountered in professional operations where hens are generally culled after their pullet year. The organism can live in the soil, so keeping the chickens' environment clean will certainly help minimise the likelihood of tuberculosis occurring, but you will usually not know that anything is wrong until you find a dead bird among your flock. Veterinary pathology testing of the dead bird can provide a diagnosis. If tuberculosis is diagnosed, you will need to thoroughly clean and disinfect the housing and equipment, and replace all litter and dirt.

Epidemics

It all sounds pretty scary, doesn't it? There seem to be so many things that can go wrong. 'Am I going to be spending all my time and money on veterinarian bills?' you're probably asking yourself. But relax for a moment. The fact is that the maintenance of a healthy, clean environment, together with the provision of a balanced diet and an adequate exercise space, will go a long way towards keeping your chickens fit and healthy. You can increase your chances of maintaining a healthy flock by keeping in contact with the poultry division of your state department of agriculture, so that you are aware of any outbreaks among poultry in your area. Quarantine any new birds and avoid mixing young birds with older stock. If you're quick off the mark and notice one or two chickens have contracted a disease, isolation and care may preserve both them and the rest of the flock. Having said that, if you are faced with a number of your flock succumbing to a particularly virulent and harmful disease, then you should seek veterinary advice early if you are to have the best chance of diagnosing and resolving the problem. Alternatively, you may choose to cull your entire flock and start all over again, as cruel and heartless as that sounds.

Egg Eating

Some hens discover how delicious eggs taste and develop a liking for them. When this occurs, you have competition on your hands. The reasons for egg eating are many and range from a lack of calcium to boredom. One way to determine if a calcium deficiency is the problem is to check the thickness of the eggshells as you collect them. Thin shells are a sure sign that a hen is not getting enough calcium and can also be the cause of her developing a taste for eggs in the first place – a yolk may have seeped through the thin shell, providing a temptation just too good to resist. If you suspect calcium deficiency, increase the calcium intake in the chickens' feed. As mentioned earlier, ground oyster shells are a superb source of calcium.

Your hens may also have gotten a taste for eggs by eating shells that you have thrown in among their scraps. To avoid this, make sure any shells are ground down and mixed with other scraps so they are unrecognisable as the items that appear in the nesting boxes. Of course, the best way to reduce or eliminate the problem of egg eating is to collect the eggs three times a day and keep the nesting area dark so that the eggs don't stand out. If you are unable to stop a chicken eating eggs, you have two options: trimming its beak (see page 101) or culling it.

Feather Pecking

Feather pecking involves chickens pulling out their own feathers or those of other chickens. At its most extreme, it can lead to cannibalism (see page 24). Feather pecking is usually a result of unsatisfactory external conditions such as overcrowding or an inadequate diet.

Changing these conditions is the best remedy, though some chicken owners trim offenders' beaks (see page 101). Feather pecking is easy to spot, being characterised by the absence of clumps of feathers and by wounds on the skin. Chickens that have been victims of severe feather pecking should be isolated until their wounds heal. Healing can be hastened by the application of an approved antiseptic treatment. Chickens that are serial offenders may have to be removed from the coop.

Wing Clipping

If you want to avoid the problem of having chickens that like to take flight, select heavy standard breeds that cannot get off the ground. But if you fancy a breed that has the ability to fly, then you are most likely going to have to learn to clip their wings. It's a simple process that involves catching the bird, stretching out one of the wings, and using a pair of scissors to cut off a few of the flight feathers near the edge of the outstretched wing (see Figure 23). The chicken's balance will be skewed and they will be unable to go far.

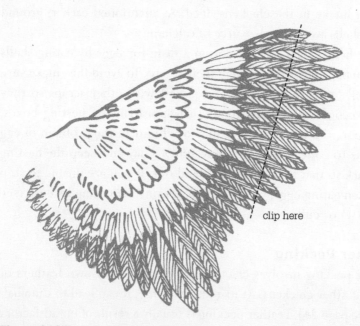

Figure 23 Where to clip the wing

If you do have a flighty breed, you are likely to become quite proficient at wing clipping because the flight feathers will grow back and you'll have to repeat the process. The best time to do so is after the birds have moulted and their feathers are growing out. Don't worry about hurting them, they don't feel a thing. Well, almost nothing. But before you go sharpening your scissors, be aware of two things: chickens with clipped wings cannot be entered in shows and they cannot fly away from predators.

Beak Trimming

Beak trimming is done to combat cannibalism, feather pecking and, to a lesser extent, egg eating. It involves using a heated blade to cut back part of the upper or lower beak. In Australia, the Welfare Code recommends that no more than half of the upper and one third of the lower beak be removed. Some chicken owners consider beak trimming to be cruel, even barbaric. It is common practice among commercial operators, usually on day-old chicks, but you should only consider it as a last resort after all other remedial options have failed. Even if you decide to have day-old chicks' beaks trimmed, you should leave the actual mechanics of the job to an expert beak trimmer. In the hands of someone inexperienced, the damage can be painful and render the chicken incapable of undertaking vital tasks.

Breeding Chickens

It's one thing to keep a few chickens for eggs; it's quite another to keep chickens to breed. That's not to say it shouldn't be done. Quite the contrary. It's a fascinating process and extremely rewarding. But it will add a lot of time to your hobby.

SETTING AND HATCHING

Most people assume that if they are going to breed their own chickens, they will need a rooster. That's not necessarily the case. Of course a rooster is required to mate with a hen to produce a fertilised egg but it is possible to buy fertilised eggs from a hatchery. However, if you wish to be involved from go to whoa, so to speak, stay tuned. Whichever option you choose, you should remember that if you want to maximise the number of eggs you get from your flock, you should replace your primary layers every two years.

The breeding season generally starts at the end of winter/start of spring and lasts until the beginning of summer. You'll need one rooster for every 10–12 hens. If you don't like the idea of having a rooster all the time, you may be able to borrow one for 2–3 weeks. This will produce ample fertilised eggs.

During the mating process, the rooster will parade around a hen, displaying its feathers and chortling loudly. The hen, for its part, indicates its interest by crouching so that the rooster can mount her. And when that's all over, and 25 hours have passed, the egg is laid.

Selecting Eggs

If you decide to go down the road of having a rooster, rather than buying fertilised eggs, then you will need to decide which eggs to keep in order to produce chicks. You don't want to end up with more chicks than you can handle.

There are two rules well worth remembering when selecting eggs. The first is that the fittest and healthiest chickens are usually the first 20–30 bred after the moulting season. The second is that you should examine the eggs very carefully and reject the ones that have flaws such as cracked or thin shells, or eggs that are abnormally large, small or misshapen. These eggs have a far lesser chance of producing healthy chicks than flawless eggs. And even if you have chosen what you consider to be perfect eggs, it is extremely unlikely that they will all hatch. You can expect a success rate of 50–75 per cent.

Hens don't necessarily incubate their own eggs. In fact, once you have selected the eggs that you consider to be the best of the bunch, you are going to have to identify the best brooder/s among your hens. In the meantime, you'll need to do something with the eggs. Store them somewhere cool and dry, where the temperature is constantly 10–15°C. Place them on their side but make sure you turn them twice a day. You can keep them in these conditions for about a week before positioning them under a hen (or, alternatively, in an artificial incubator – see page 108).

When you handle fertilised eggs, do so very carefully. In fact, minimise the handling as much as possible. Before picking them up, always wash your hands with water and a mild soap, and dry them well. This will prevent the natural oils in your skin from passing onto the egg and clogging up the minute holes that allow oxygen and carbon dioxide to pass through.

Selecting a Hen

You might think that it will be easy to find a suitable hen to sit on your eggs. After all, that's what they're meant to do. In one way you're right. Originally, in the wild, hens were naturally broody. But centuries of domesticating chickens has reduced the brooding instinct in many of the breeds.

If you keep bantams and have a Silkie bantam hen, you should have no problems at all. They are great brooders and individual Silkie hens are often sold to bantam keepers for just this purpose. Similarly, Orpington and Sussex hens haven't forgotten what they're supposed to do. On the other hand, the focus on getting Leghorn hens to continually produce eggs for the commercial market has created a breed for whom motherhood is a very foreign concept.

Some people buy or borrow a hen that has proven to be a good brooder. Such action can be successful but is not always the perfect solution. Firstly, it can seriously disrupt the pecking order within the existing flock, upsetting both the new arrival and the permanent residents. Secondly, the new environment may not suit the hen. A hen that is broody in one place won't automatically be broody elsewhere. Which puts you back where you started – trying to identify a brooder among your existing hens.

Identifying a brooder among your hens may take a bit of work, but you'll get there eventually. The easiest strategy is to put your hens to the test. For the first test, place some fake eggs in a dedicated nesting spot – you should use fake eggs because you don't want real ones cracked or pushed out of the nest onto the floor. Once you have put the fake eggs in the nest, place a hen on top of them. The best time to do this is at night when your hen is roosting, or about to roost. Hopefully, when morning comes, your hen is still on the nest. If not, try another hen as this one obviously does not have the maternal instinct you're after. Once you've found a hen that is at least staying on the nest, it's time for the next stage of their exam.

The next stage involves trying to 'steal' one of the fake eggs from underneath a hen. If the hen is nonchalant about your actions and allows you to take it away, she's failed the test and it's back to the drawing board. But if she ruffles her feathers, pecks at you and looks as if she's ready to fight to the death to protect the eggs, you've got yourself a brooder.

Making a Nest

So, you've got your fertilised eggs and you've found your brooder. You're halfway there. Now you've got to make sure that the spot you've chosen for the setting and hatching processes meets particular requirements.

The nesting area should be dark and well ventilated. Your hen is going to be in there for about three weeks, leaving the nest only occasionally. Make sure that your other chickens can't get into the nesting area, lest they attack your hen or start eating the eggs. And speaking of attacks, you should also put up some wire or netting to keep rats, mice and snakes from getting in. If you have more than one hen setting at the same time, you can put them within the same area but give them plenty of space.

The actual nest in which the hen is going to set must be big enough to allow the hen to move around and turn her eggs. Hens turn their eggs up to 100 times a day. The reason that eggs are turned is to prevent the developing embryo sticking to the lining of the shell. There should be plenty of litter in the nest for comfort and warmth. Use cedar shavings as litter if you can, to minimise the presence of mites and other parasites. The nest should be well rounded so that eggs don't roll out of the nest. The surrounding area should have litter or some other soft padding strewn about so that when the eggs hatch, the chicks are safe. Water and feed should be positioned relatively close to the nest, but far enough away that the hen has to get out and stretch her legs. As a final touch, sprinkle some approved anti-parasite powder in and around the nest.

Incubation

Now you've got your hen, your eggs and your nest. It's time to incubate the eggs (see Figure 24). The incubation period is 21 days. Before allowing your hen to get too comfortable, check her thoroughly for mites and lice. While she is setting, she will not be able to take a dust bath and a mite or lice infection will leave her endlessly scratching and extremely uncomfortable and frustrated.

There's not a lot you have to do during the incubation period, except keep an eye on the hen and eggs to make sure that everything seems okay. Make sure that your hen leaves her nest a couple of times a day for food and water. This doesn't only benefit the hen, it gives the eggs an opportunity to cool down a little bit. But you don't want your hen staying off the nest for more than 30 minutes at a time. While the hen is away from the nest, check the eggs for dirt and droppings, and carefully wipe off any offending materials with a dry cloth. Be warned, however, that your hen may be very protective of her eggs.

Sometimes, a hen will have such a brooding instinct that she will put the welfare of her eggs well above her own. She may not even leave her nest for food and water. In such cases, you are going to have to take the food and water to her. Make sure that you feed her grains, rather than mash, so that her droppings are solid and will not soil the eggs too much.

All being well, on the twenty-first day of setting, the chicks will start appearing. The day before, you'll notice a hole starting to appear in the shell. The process whereby the chick pecks its way out of the shell is called pipping. Most chicks emerge without the need for human intervention, but if you notice a chick having problems getting out of its shell, you can enlarge the hole and spray a small amount of warm water over the chick (this should be carried out as a last resort only). In general, if you leave them alone the chick and the hen will complete the job themselves.

Figure 24 A hen setting (sitting on her eggs)

Candling

Candling is a way to check whether the eggs being set are fertile. Obviously you can let the hen sit on all the eggs and just wait and see what happens after 21 days. However, it's best to get rid of the infertile ones because they will start to rot and let off gases that could contaminate your fertile eggs. Candling takes place after the hens have been setting for one week. You will need a 75-watt light globe and a cardboard box big enough to go over the top of the globe. (The term candling came about because traditionally, the process involved the use of a candle, rather than a light globe.) Make a hole of about 2.5 centimetres diameter in the box and place the box over the light globe while it's connected to a power source. Take the eggs from the nest one at a time, and hold them above or alongside the light coming from the hole in the box. It's obviously best to do this at night or in a dark spot. A fertile egg will have a dark blob in it, while an infertile one will be clear.

ARTIFICIAL INCUBATORS

An artificial incubator is a device that replicates the conditions an egg encounters when ensconced under a hen. It keeps an egg warm and moist, and after 21 days a chick should emerge. There are several reasons why you might use an incubator, including having sick or unreliable hens, or too few hens for the number of fertilised eggs you have. If you are really adventurous, you could breed the very first members of your flock this way. You would have to buy fertilised eggs from a hatchery and then incubate them. However, it's certainly not the easiest way to start a flock.

Forced Air and Still Air

The first thing you need to know about incubators is that there are many different types and models, all with their own very specific instructions. You must follow the instructions for your incubator exactly, or risk ending up with nothing to show for your time and effort. Some people have been known to build their own incubators but this isn't recommended unless you have formidable building and science skills and knowledge. Small variations in temperature and humidity conditions can be disastrous.

There are incubators that fit just half a dozen eggs and there are incubators that can hold thousands of eggs. There are incubators that sit on tables or bench tops, and there are free-standing cabinet incubators that have rows and rows of trays. Some of the smaller ones have glass tops or sides so that you can observe what's happening inside. If you think that curiosity is going to get the better of you, it's probably a good idea to get one of these, as continually opening and closing the incubator door can reduce the temperature of the eggs.

Despite there being so many models and sizes, there are essentially two types of incubators: forced air and still air. Forced air incubators have a fan system that circulates the air, while still air incubators have a vent system, with cold air entering at the bottom and warm air escaping through the upper vents. Forced air incubators are more popular and are generally more reliable.

Positioning the Incubator

Place your incubator in a room that has a constant temperature of 20–25°C. Do not place it near a window, heater or direct sunlight. The temperature you set the incubator at will be indicated in your instructions, and differs for forced air and still air incubators. You should set your incubator at the instructed temperature 2 days before placing the eggs inside. Once the eggs are incubating, you must check the temperature several times a day to make sure all is okay. It's also a good idea to have an emergency generator standing by just in case you experience a power failure. Apart from checking the temperature, make sure the water pan is always topped up, so the humidity remains at the right level.

Eggs in the Incubator

When you place the eggs in the incubator, make sure they are all clean. Dirty eggs carry bacteria that can infect the entire batch being incubated. Wipe all dirt off the eggs with a dry cloth before placing them inside. Eggs should be placed in their tray with the small end pointed slightly downward. You want the pipping to occur at the large end where the air cell is.

Some incubators have automatic egg turners that turn the eggs every hour. If you don't have one of these, you are going to have to do this job yourself. No-one expects you to turn all the eggs hourly, but you should make sure that all eggs are turned at least three times a day. To turn the eggs, roll them with the palm of your hand at evenly spaced intervals. In other words, don't turn them 180° one time, and 90° the next time. It's useful to devise a method of knowing which eggs you've turned and how far you've turned them. Many breeders use a pencil to mark the eggs in a way that provides them with such information.

Once the eggs have been in the incubator for a week, it's time for the candling process. It's exactly the same as for the naturally incubating eggs (see page 106). Remove any infertile eggs from the incubator. As day 21 approaches, get ready for hatching. About 24 hours before the chicks hatch, you will notice the pipping beginning. If you have trays of chickens in a cabinet incubator, place the eggs that are the most advanced toward the bottom so that waste and down do not fall on the eggs below. Regularly clean the water tray and the floor to rid the incubator of broken shell and loose down.

Hatching Time

When the chicks have hatched, leave them in the incubator until they have fluffed out. This can take 12–24 hours. You don't have to feed the chicks during this time; in fact, you can go without feeding them for up to 48 hours, because they are quite happily absorbing the yolk that was in the egg with them. As with natural incubation, don't be disappointed if you don't have 100 per cent success. In fact, you should be amazed if you do. But one thing's for certain: you'll start bonding with your fluffy new companions almost immediately.

THE FIRST 6 WEEKS

If your chicks were hatched by a hen, then you really just have to keep an eye on them to make sure that the hen remains attentive to, and interested in, the new arrivals. Provide the chicks with the appropriate feed and regular supplies of cool, fresh water. Most importantly, ensure that the hen and the chickens remain in an enclosure that is secure from the other chickens and predators. By about six weeks, the chicks will be ready to fend for themselves and can be introduced to the rest of the flock. By this time, the hen will start laying again and will lose interest in the chicks anyway.

However, it's a completely different story if your chicks were hatched in an incubator, you bought some day-old chicks, or your hen's maternal instincts desert her shortly after hatching. In these cases, you are going to need to house your chicks in a brooder.

The Brooder

A brooder is a wire cage or a ventilated box in which chicks spend the first few weeks of their life, receiving the nourishment and warmth they require to become healthy chickens. You can buy a commercial brooder if you like but you can also use a wire cage (such as a bird cage), a cardboard box or even an old fish tank.

To provide the necessary heat, hang a light globe above the brooder. Make sure that you can move the light up or down to create the right temperature. During the first week, the temperature should be about 35–38°C, and should then be reduced by 3°C each week. Raising the light about 6–8 centimetres each time should produce the necessary reduction in temperature. The best way to check whether the temperature is correct is to observe where your chicks are congregating. If they huddle under the light, it's too cold for them; if they seek sanctuary around the edges of the brooder, it's too hot; and if they're pretty evenly spread within the brooder then it's just right.

The brooder does not go in the coop with your other chickens. It needs to be placed somewhere in your house where you can keep a regular watch on the welfare of the chicks. Remember, though, that chicks expel a lot of waste and a brooder can emit a strong odour.

Also, chicks cheep a lot. In fact, they cheep incessantly. So placing a brooder in your bedroom will not be conducive to getting a good night's sleep.

Inside the brooder you need to place some litter (newspaper) on the base, a feeder and a waterer, and some rags. Don't use wood shavings as litter because the chicks may try to eat them. Change the soiled newspaper at least once a day.

The feeders and waterers can be bought from your poultry supplier, as can the special high-protein feed. You should also get some chicken grit to sprinkle over the feed. The waterer should be cleaned and replenished at least once a day, preferably twice. Water-bound bacteria is the most likely source of disease in these first crucial weeks. Also, make sure the water is cool – ensure this by positioning the waterer so that it is not directly under the light.

Place some cardboard around the outside of the brooder to keep drafts off the chicks. This is a bit of a juggling act because you also want to keep the brooder in a place where it will receive fresh air.

Place clean rags on the base of the brooder for the chicks to sleep on. They don't roost for the first three weeks or so, they just collapse on the spot when they're tired. Many first-time chicken keepers have panicked at this sight, thinking they've stumbled upon a deadly epidemic of mammoth proportions. Just make sure that the rags don't have loose threads on them or you may face such a disaster. Chicks will try eating anything.

With careful management, and some luck, you've given your chicks a great start in the first week or so of their life. The only other unexpected stumbling block you should be aware of is a condition known as pasting up. This occurs when a chick's droppings stick to its vent until new droppings cannot be expelled. If you see this occurring, pick the dried droppings off and wipe the area with a damp, warm cloth. The chick might object, but believe me, you're doing it a great favour.

After about 3 weeks, put a small perch in the brooder. A piece of dowel is adequate. This is to teach the chicks to roost. Give them a day or two to check it out and see if they use it as a roost. If not, gently pick them up and put them on the perch. They'll quickly get the idea.

After about 4 weeks, you can take the chicks out of the brooder and place them in a larger pen. By this time they'll be getting too big for the brooder anyway. Gradually give them some 'supervised' time in the coop, making sure that they are not accosted by other chickens. By the time they are full-feathered, they should be ready for life in the coop.

Sexing

Sexing is the term given to the process of identifying gender in chickens. It's something that you should do during the first six weeks or so. Professional chicken sexers can differentiate between male and female chicks at the day-old stage, which is why it's possible to buy a batch of female day-old chicks. However, the professional method involves an internal examination that is particularly fiddly. (A chick's vent is opened wide enough to reveal the internal genital area, in which is a tiny bead-like organ. The curvature of this bead enables the professionals to make their call.) You are far better off waiting a few weeks for some more obvious signs to appear.

The two easiest characteristics to identify are the size and colour of the comb, and the presence of tail feathers. At around six weeks, male chicks have larger and redder combs. Their tail feathers also arrive earlier. Remember, cockerels grow up to be roosters and you may not want (or be allowed) to keep roosters.

DEALING WITH ROOSTERS

Let's say you want to keep roosters to mate with your hens, and your local council (and neighbours) are more than happy for you to do so. You need one rooster for approximately every 10 hens. Just as the stereotype suggests, roosters do strut and they spend a great deal of time trying to impress the hens. If you have more than one rooster, there may be fights, but again this is largely for show. Rarely, if ever, will you have to separate them. And you'll know when you do – the blood and shrieks will indicate that this is a world title bout rather than a bit of friendly banter.

The main thing to be aware of concerning roosters and the 'mating game' is that roosters can cause serious pain to hens with their claws and spurs. Their claws can be clipped to minimise this happening but if it does occur more than once to the same hen, it's best to separate her from that particular rooster. Not surprisingly, being repeatedly clawed and spurred during mating could put her off the whole process.

And remember, it's best not to get too attached to a rooster if you're keeping him purely for breeding purposes. You can only expect about two years of good, fertile service from a rooster.

Chickens for Show

Keeping chickens for show is a completely different affair to keeping them for their eggs or as pets. And if you are going to show your chickens, you need to make this decision at the very beginning, before you even purchase your first chicks.

There are very definite guidelines surrounding the showing of chickens. These are called standards and they differ from country to country. In Australia, the physical characteristics that a chicken has to conform to in order to be shown are contained in the *Australian Poultry Standards*. These standards can be both exacting and brutal. They do not allow for any flaws. They include everything from the desired weight and height of the birds to the colour of their plumage, as well as less obvious features such as the number of toes they have.

Anyone considering showing chickens should join an association specialising in the breed they are going to keep. The members of this association can provide invaluable advice and information, and are always willing to do so. Magazines and online newsletters are also good sources of information. Remember, there are many people out there with years of experience you can harness. There's no point reinventing the wheel. There's also no point trying to radically overturn the existing order. The exhibiting 'scene' can fairly be described as conservative, and the rules and regulations are there for a reason – to maintain the characteristics in a breed that are regarded as ideal.

Alongside joining an association and subscribing to magazines and newsletters, you should also attend as many shows as possible. You'll soon find out exactly what you (and your chickens) are in for.

When buying stock for show, rather than for eggs or pets, there is a distinct advantage in buying mature stock, rather than chicks. After all, who knows what those cute little chicks are going to grow up to look like? If you do purchase younger birds direct from the breeder, ask to see their parents. It's no guarantee but it will provide some sort of guide as to what your potential show bird will look like when it matures. Beware though that some breeders are so keen to replicate the prize-winning traits of their birds that they continually breed within a successful family, resulting in chickens that are so inbred that they develop low resistance to diseases. Being susceptible to diseases is bad enough if a chicken is confined to its coop, but it can be fatal on the show circuit, where hundreds of chickens are kept in very close quarters. Shows can be a great spreader of diseases.

Let's say you've taken the plunge and have reared some 'flawless' chickens that are ready to show. Don't get too dispirited if success is not immediate. Like anything, it's a learning process. The most important thing is to find out what the difference is between your chickens and those that end up with the rosettes or the prize money. Feel free to ask judges why they made certain decisions, though do so with grace, not as a sore loser. The judges will be more than happy to help a novice. Learn from what they say and put it into practice, either in preparing your chickens or when purchasing them.

Once you've won a prize or two, you might even consider breeding from your chickens and making some money out of your hobby. After all, the offspring of show champions are highly sought. Before doing so, consider the difficulties in replicating your birds. The genetic possibilities are numerous and in the world of chickens, like doesn't always produce like. With some breeds, the offspring of a rooster with feathers of a particular colour and a similarly coloured hen have only a 25 per cent chance of having the same colouring. If that colour is the competition standard, three out four of the offspring are likely to be of no value whatsoever.

There are hundreds of shows held around Australia each year, some of which are specifically for chickens, others which have chickens among many other categories. The best known and most prestigious of the latter are the Royal Shows held in many capital cities. The shows are usually structured in the same way: there are competitions for each variety, then these winners compete for best in breed, and finally the best in breed winners compete for overall champion.

The final piece of advice for someone thinking of keeping chickens for show is to make sure that the breed and variety that you are planning to rear and show can be entered in competitions. While there are many, many categories at most of the shows, not every breed and variety are catered for. The Royal Shows tend to have the widest range and as a guide, here is a list of the breeds and varieties catered for at the 2007 Sydney Royal Easter Show. If your preferred breed and variety is not included in this list, consider one of the ones that is. Note that most of these categories have separate competitions for standards and bantams.

- Ancona Red
- Ancona, rose comb
- Ancona, single comb
- Andalusian
- Araucana, any colour
- Australian Game, black-red
- Australian Game, duck-wing
- Australian Game, pile or white
- Australian Game, any other colour
- Australian Henfeather Pit
- Australian Muff Pit Game
- Australian Pit Game
- Australorp, black
- Australorp, any other colour
- Barnevelder, double laced
- Barnevelder, any other colour
- Belgian Barbu d'Anvers, any colour
- Belgian Barbu d'Uccle, black
- Belgian Barbu d'Uccle, white
- Belgian Barbu d'Uccle, any other colour
- Belgian Barbu d'Watermael, any colour
- Brahma, light
- Brahma, any other colour
- Campbell, any colour
- Campine
- Faverolle, salmon
- Faverolle, any other colour
- Frizzle, white
- Frizzle, any other colour
- Hamburg, black
- Hamburg, gold pencilled
- Hamburg, silver spangled
- Hamburg, any other colour
- Indian Game, jubilee or blue
- Indian Game, dark
- Indian Game, any other colour
- Indian Runner, any colour
- Japanese, white

- Japanese, any other colour
- Langshan Australian, black
- Langshan Australian, blue
- Langshan Australian, white
- Langshan Australian, any other colour
- Langshan Croad
- Leghorn, black
- Leghorn, blue
- Leghorn, brown
- Leghorn, buff
- Leghorn, white
- Leghorn, any other colour
- Minorca, black
- Modern Game, birchen
- Modern Game, black-red
- Modern Game, blue-red
- Modern Game, brown-red
- Modern Game, duck wing,
- Modern Game, pile
- Modern Game, wheaten
- Modern Game, any other colour
- Muscovy, any colour
- New Hampshire
- Old English Game, black
- Old English Game, black-red,
- Old English Game, blue
- Old English Game, blue-red
- Old English Game, blue-tailed
- Old English Game, brown-red
- Old English Game, clay
- Old English Game, dark grey
- Old English Game, duck-wing
- Old English Game, ginger
- Old English Game, golden
- Old English Game, partridge or wheaten
- Old English Game, pile
- Old English Game, spangled
- Old English Game, any other colour
- Orpington, black
- Orpington, blue
- Orpington, buff
- Orpington, any other colour
- Pekin, birchen or brown-red
- Pekin, black
- Pekin, blue
- Pekin, buff
- Pekin, mottle
- Pekin, partridge
- Pekin, white
- Pekin, any other colour
- Plymouth Rock, dark barred
- Plymouth Rock, light barred
- Plymouth Rock, any other colour
- Polish, any colour
- Rhode Island Red
- Rhode Island Red, rose comb
- Rhode Island White
- Sebright, golden
- Sebright, silver
- Sebright, any other colour
- Silkie, bearded

- Silkie, black
- Silkie, buff
- Silkie, partridge
- Silkie, white
- Silkie, any other colour
- Spanish
- Sussex, light
- Sussex, any other colour
- Welsummer, any colour
- Wyandotte, buff
- Wyandotte, Columbian
- Wyandotte, gold laced
- Wyandotte, partridge
- Wyandotte, silver laced
- Wyandotte, silver pencilled
- Wyandotte, white
- Wyandotte, any other colour

Resources and Suppliers

STATE AND TERRITORY GOVERNMENT DEPARTMENTS

Australian Capital Territory Department of Territory and Municipal Services
Macarthur House, 12 Wattle Street
Lyneham ACT 2602
ph 13 22 81
e   info@tams.act.gov.au
w   www.tams.act.gov.au

New South Wales Department of Primary Industries
161 Kite Street
Orange NSW 2800
ph (02) 6391 3100
e   nsw.agriculture@agric.nsw.gov.au
w   www.dpi.nsw.gov.au

Northern Territory Department of Primary Industry, Fisheries and Mines
Berrimah Farm, Makagon Road
Berrimah NT 0828
ph (08) 8999 5511
e   info.DPIFM@nt.gov.au
w   www.nt.gov.au/dpifm

Queensland Department of Primary Industries
80 Ann Street
Brisbane Qld 4000
ph 13 25 23
e   callweb@dpi.qld.gov.au
w   www.dpi.qld.gov.au

South Australian Department of Primary Industries and Resources
33 Flemington Street
Glenside SA 5064
ph (08) 8207 7842
e   pirsa.livestock@saugov.sa.gov.au
w   www.pir.sa.gov.au

Tasmanian Department of Primary Industries and Water
1 Franklin Wharf
Hobart Tas. 7000
ph 1300 368 550
e   info@dpiw.tas.gov.au
w   www.dpiw.tas.gov.au

Victorian Department
of Primary Industries
1 Spring Street
Melbourne Vic. 3000
ph 136 186
e   customer.service@dpi.vic.
gov.au
w www.dpi.vic.gov.au

Western Australian
Department of Agriculture
and Food
3 Baron-Hay Court
South Perth WA 6151
ph (08) 9368 3333
e   enquiries@agric.wa.gov.au
w www.agric.wa.gov.au

STATE POULTRY ASSOCIATIONS

You can contact each of the organisations listed for a full list of their affiliated associations.

Feather Clubs Association of Queensland
c/o David Simons (Secretary)
294 Old Toorbul Point Road
Caboolture Qld 4510
ph (07) 5499 0553
e   fcaqi@ezweb.com.au
w fcaqi.tripod.com

Victorian Poultry Fanciers Association
c/o Mr R Lindsay (Secretary)
15 York Street
Linton Vic. 3360
ph (03) 5344 7284
w www.vpfa.org.au

South Australian Poultry Association
c/o Mary Scruby (Treasurer)
34 Balmoral Avenue
North Brighton SA 5048
ph (08) 8296 9535
e   maryscruby@ozemail.com.au
w www.sapoultryassoc.org.au

CLUBS AND ASSOCIATIONS (BY BREED)

Note: the following are national associations. For local associations promoting your breed/s, contact the national associations with your details.

Australorp Club of Australia
c/o Ross Summerell (Secretary)
PO Box 70
Tamborine Qld 4270
e   austclub@tpg.com.au
w   www.australorps.com

Belgian Bantam Club of Australia
c/o Irene Hannan (Secretary);
'Mountain View', 930 Caparra Road
Caparra via Wingham NSW 2429
e   channan@tpgi.com.au
w   users.tpg.com.au/channan/index.html

Brahma Club of Australia
c/o The Secretary
PO Box 100
Maldon Vic. 3463
e   brahma2000@bonbon.net
w   www.geocities.com/brahma2000

Faverolles Club of Australia
c/o Irene Hannan (Secretary)
ph (02) 6550 7295

Langshan Club of Australia
c/o Kim Birchall-Blayney (Secretary)
ph (02) 9580 2895

Leghorn Club of Australia
c/o Ken Bergin (Secretary)
PO Box 5
Summer Hill NSW 2130
ph (02) 9799 6283

Old English Game Fowl Club of Australia
c/o Tony Davis (Secretary)
55 Appletree Road
Holmesville NSW 2286
ph (02) 4953 3260

Orpington Club of Australia
c/o Judy Witney (Secretary)
94 Allen Road East
Lardner Vic. 3821
e   witney@dcsi.net.au
w   www.orpington.backyardpoultry.com

Australian National Pekin Club
c/o David Plant (Secretary)
46 Newcastle Street
East Maitland NSW 2323
ph (02) 4933 0733
e   pekinsaust@bigpond.com
w   www.pekinclub.backyardpoultry.com

Plymouth Rock Club
of Australia
c/o Adrian Kuys (Secretary)
ph (08) 8836 7242

Crested Breeds Club of
Australia (Polish)
c/o Kory Chapman (Secretary)
PO Box 738
Warwick Qld 4370
e   shadow_valley@yahoo.com

Silkie Club of Australia
c/o The Secretary
31 Kerry Road
Blacktown NSW 2148

Wyandotte Club of Australia
c/o Duane Rhall (Secretary)
PO Box 307
Parkes NSW 2870
e   whitewyandottes@ozemail.
     com.au
w members.ozemail.com.
     au/~whitewyandottes

BREEDERS

There are hundreds of breeders around the country. To find your nearest breeder, contact the relevant national association as listed in the previous section or use one of the following online directories that enable searches by state and/or breed.

Australian and New Zealand
Poultry Breeders Directory
w www.backyardpoultry.com/
     breeder

Australian Online Poultry
Breeders Directory
w insight.iinet.net.au/breeders

Oz Breeders
w www.ozbreeders.com

Oz Fanciers
w home.iprimus.com.au/onslo/
     ozfanciers.html

RESOURCES AND SUPPLIERS

EQUIPMENT AND FEED SUPPLIERS

New South Wales

Agritech Australia
9 Crusader Road
Galston NSW 2159
ph (02) 9653 3344

Condell Park Produce
Rear 44 Simmat Avenue
Condell Park NSW 2200
ph (02) 9790 6231
e   enquiries@cpproduce.com.au
w   www.cpproduce.com.au

The Discerning Gourmet
Unit 2/5 Tathra Street
Gosford West NSW 2250
ph (02) 4322 5180

Dunogan Farm Tech Pty Ltd
8 Bellevue Crescent
Tamworth NSW 2340
ph (02) 6766 9909
e   info@dunoganfarmtech.
     com.au
w   www.dunoganfarmtech.
     com.au

Dutts Engineering Works
4 Isabel Street
Cecil Hills NSW 2171
ph 0407 237 403

Herfo's Shell Grit
342 Wisemans Ferry Road
Central Mangrove NSW 2250
ph (02) 4373 1316

H J N International
4 Coates Place
Wetherill Park NSW 2164
ph (02) 9604 6466
w   www.hjninternational.com.au

Imexco Australia Pty Ltd
Winta Road
Tea Gardens NSW 2324
ph (02) 4997 2045

Kensington Produce Pty Ltd
2/19a Baker Street
Banksmeadow NSW 2019
ph (02) 9666 7755

Linco Food Systems Pty Ltd
49 Prince William Drive
Seven Hills NSW 2147
ph (02) 9624 2055

Multiquip Pty Ltd
Corner Tenth Avenue & Kelly
Street
Austral NSW 2171
ph (02) 9606 9011
e   info@multiquip.com.au
w   www.multiquip.com.au

Patarker Pty Ltd
81 Station Road
Seven Hills NSW   2147
ph   (02) 9838 7980

Poultry Processors
186 Hume Highway
Lansvale NSW   2166
ph   0414 551 191

Procut Australia Pty Ltd
2/10 Amour Street
Milperra NSW   2214
ph   (02) 9792 7700

Roden Poultry Service
7 Wendy Place
Toongabbie NSW   2146
ph   0407 107 336

Tec Sys Animal Management
5 Doomben Avenue
Eastwood NSW   2122
ph   (02) 9896 3455

Universal Processing
Equipment Co Pty Ltd
40 Flora Street
Kirrawee NSW   2232
ph   (02) 9542 1611

YCC Poultry
25 Warren Avenue
Bankstown NSW   2200
ph   (02) 9790 6474

Queensland

Allen's Rural & Hardware
Supplies
11 Kingsthorpe-Haden Road
Kingsthorpe Qld   4400
ph   (07) 4630 0179

Bagley Produce
7 River Street
Mackay Qld   4740
ph   (07) 4951 4911

Ballandean General Store
New England Highway
Ballandean Qld   4382
ph   (07) 4684 1103

Boylans Produce
29 Woondooma Street
Bundaberg Qld   4670
ph   (07) 4151 3826

Dillon & Co General Produce
Mt Lindesay Highway (Corner
St Aldwyn Road)
North Maclean Qld   4280
ph   (07) 3802 1955

Fearnside & Co
18 Davadi Street
Stanthorpe Qld   4380
ph   (07) 4681 1173

Highfields Produce, Hardware
& Veterinary Supplies
New England Highway
Highfields Qld   4352
ph (07) 4630 8297

Landline Industries Pty Ltd
14 Chambers Road
Woodford Qld   4514
ph (07) 5496 3311

McCallum Made Chicken
Tractors
690 Old Goombooran Road
Gympie Qld   4570
ph (07) 5483 3800

Morayfield Produce & Rural
Supplies
166 Morayfield Road
Morayfield Qld   4506
ph (07) 5495 5444

Nerang Park Poultry
45 Gilston Road
Nerang Qld   4211
ph (07) 5578 1666

Outback Environmental
Controls Pty Ltd
2/32 Billabong Street
Stafford Qld   4053
ph (07) 3352 6677

Parkhurst Produce
34 Hollingsworth Street
Kawana Qld   4701
ph (07) 4932 7066

Plucker Industries
138 Walli Creek Road
Kenilworth Qld   4574
ph (07) 5472 3182

Poultry Livestock Service
& Incubators
3 Duffield Road
Kallangur Qld   4503
ph (07) 3204 4077

Range Harvester Pty Ltd
5 Industrial Avenue
Caloundra West Qld   4551
ph (07) 5437 0275

Redmond E M & Co (Gatton)
Pty Ltd
Crescent Street
Gatton Qld   4343
ph (07) 5462 1139

Specialised Farm Services
Pty Ltd
563 D'Arcy Road
Carina Qld   4152
ph 0408 886 862

Spitwater Queensland
1909 Ipswich Road
Rocklea Qld   4106
ph 1300 880 403

Sunridge Poultry Farm
Bruce Highway
Forest Glen Qld   4556
ph (07) 5445 1647

South Australia

Clay & Mineral
Sales – Shellgrit Supplies
Lot 1 Hancock Road
Golden Grove SA   5125
ph (08) 8251 4000

Gilbertson's Fodder Store
570 Main North Road
Gepps Cross SA   5094
ph (08) 8262 1154

Incubators & More Pty Ltd
212a Gouger Street
Adelaide SA   5000
ph (08) 8231 8778

Intensive Farming Supplies
Australia
Unit 4/9 Cardiff Court
Cavan SA   5094
ph (08) 8349 8077
e   info@ifsupplies.com.au
w   www.ifsaustralia.com.au

Inverarity Permaculture
Stirling SA   5152
ph (08) 8339 3840
e   info@i-permaculture.com.au
w   www.i-permaculture.com.au

Keelan Grain & Fodder Store
Rear 291 Payneham Road
Royston Park SA   5070
ph (08) 8362 4178

Magill Grain Store
574 Magill Road
Magill SA   5072
ph (08) 8331 8159

Olivers Grain & Garden
404 Morphett Road
Warradale SA   5046
ph (08) 8377 0223

Rapuano's Grain Store
157 Trimmer Parade
Seaton SA   5023
ph (08) 8445 8919

Saleyard Fodder
173 Thomas Street
Murray Bridge SA   5253
ph (08) 8532 3030

Sanders Bros Grain & Fodder
Corner 22nd & 23rd Street
Gawler South SA   5118
ph (08) 8522 4450

Titan Poultry Equipment
12–14 Kenworth Road
Gepps Cross SA   5094
ph (08) 8359 1588

Tasmania

Fehlberg's Produce
59 Ferguson Street
Brighton Tas. 7030
ph (03) 6268 1285

Longford Chaff & Grain
144 Marlborough Street
Longford Tas. 7301
ph (03) 6391 1374

Victoria

A Better Shed
59 Slater Parade
Keilor East Vic. 3033
ph (03) 9336 3136

Acmite Shellgrit Supplies
PO Box 156
Ferny Creek Vic. 3786
ph (03) 9755 2015

Bellsouth Pty Ltd
8/7 Vesper Drive
Narre Warren Vic. 3805
ph (03) 9796 7044

Bush's Produce Stores
94 Williamson Street
Bendigo Vic. 3550
ph (03) 5443 5960

CEPA Farm Supplies
4–10 Stanley Street
Bendigo Vic. 3550
ph (03) 5443 6072

Control Medications Pty Ltd
2 Glendale Avenue
Hastings Vic. 3915
ph (03) 5979 4488

Eltham Stockfeeds
1142 Main Road
Eltham Vic. 3095
ph (03) 9439 6716

Gardenland Landscape Supplies
324 San Mateo Avenue
Mildura Vic. 3500
ph (03) 5021 3013

Gippsland Grain Stores
21 Wood Street
Bairnsdale Vic. 3875
ph (03) 5152 3093

Greensborough Grain Store
Rear Main Street
Greensborough Vic. 3088
ph (03) 9435 1339

Highlands Produce Store
125 Northern Highway
Kilmore Vic. 3764
ph (03) 5782 2166

Lloyd Trading
37–39 Seventh Street East
Mildura East Vic. 3500
ph (03) 5023 5533

McMahon J & Sons
Shell Road Point
Lonsdale Vic. 3225
ph (03) 5258 1267

OEC Southern
14 Ash Street
Leopold Vic. 3224
ph (03) 5250 3781

Patch's Canvas Manufacturing Pty Ltd
115 Hattam Street
Bendigo Vic. 3550
ph (03) 5442 3211

Peter Gibbs Stockfeeds Pty Ltd
Hartington Street
Glenroy Vic. 3046
ph (03) 9300 2088

Western Australia

Altona Hatchery Pty Ltd
344 Hawtin Road
Forrestfield WA 6058
ph (08) 9453 6611

Austral Hatchery
38 Blundell Street
West Swan WA 6055
ph (08) 9274 2670

Metrowest Automation & Control Pty Ltd
22 Claude Street
Burswood WA 6100
ph (08) 9416 0666

Picton Produce
Lot 3 South Western Highway
Picton WA 6229
ph (08) 9725 4219

Stockyard Stock Supplies
91 Pass Street
Geraldton WA 6530
ph (08) 9921 3702

U Can Hatch Us
PO Box 76
Serpentine WA 6125
ph 0417 172 013

WA Poultry Equipment
1170 Baldivis Road
Baldivis WA 6171
ph (08) 9524 1251

Wesfan Poultry Equipment
31 Morgan Street
Cannington WA 6107
ph (08) 9356 8466

WEBSITES WITH IMAGES OF BREEDS

The websites listed here each have photographs of a range of different breeds. The websites of clubs and breeders associations (see page 122) will often have photographs of specific breeds.

FeatherSite
www.feathersite.com/Poultry/BRKPoultryPage.html#Chickens

Omlet
www.omlet.co.uk/breeds/breeds.php?breed_type=Chickens

Department of Animal Science, Oklahoma State University
www.ansi.okstate.edu/poultry

My Pet Chicken
www.mypetchicken.com/chickenPics.aspx

Glossary

bantam	A small chicken (about a quarter the normal size of its breed). While the term originally referred to chickens of particular (small) breeds, it has come to refer to small-sized chickens from any breed.
broody	Describes the mood of a hen when she wants to sit on a batch of eggs.
cannibalism	The propensity for chickens to peck each other until damage has been caused.
chick	A baby chicken up to a few weeks old.
cock	A mature male chicken (also known as a rooster).
cockerel	A male chicken less than one year old.
comb	The red fleshy growth on a chicken's head.
crop	The section of a chicken's gullet where feed is softened before passing through the rest of the digestive system.
crop bound	A condition in which the crop becomes blocked with grass and/or other materials.
cull	To remove a bird that is non-productive or troublesome.
debeak	To remove part of a bird's beak in order to prevent cannibalism.
deep litter	A thick layer of straw, sawdust or shavings on the floor of the fowl house.
dust bath	An area of dry dirt where chickens roll so as to get rid of lice and mites.
egg bound	A condition that occurs when a hen is unable to pass an egg through the oviduct.
embryo	The developing chicken inside the egg.
flighty	Excitable.

free range	Describes fowl that are not enclosed during the day and are allowed to roam at will.
gizzard	An organ of the body in which food is ground with grit.
grit	Stony matter eaten by fowl to aid the grinding process.
hen	A mature female chicken.
hybrid	Offspring produced by crossing a hen and a cock of different breeds.
incubator	A machine used to artificially hatch eggs.
keel	The fowl's breastbone.
moult	To shed and replace feathers (occurs annually).
muffs	Thick tufts of feathers around the neck.
pecking order	The social order within a group of chickens.
pipping	The process whereby the chick pecks its way out of the shell.
point of lay	Pullets that have just reached the age of laying (approximately 20 weeks).
pullet	A female chicken in its first year of laying.
pure breed	A breed that has not been crossed with any other breed.
rooster	A mature male chicken (also known as a cock).
scaly leg	A disease caused by the scaly leg mite that affects the legs of fowl.
scratch	A small snack of whole or cracked grain fed to chickens in addition to their normal feed.
set	To sit on a clutch of eggs, to incubate them.
sexing	The act of determining the sex of a chicken.
standard	The considered 'ideal' size and characteristics of a particular breed.
vent	A chicken's orifice, through which both faeces and eggs come out.
wattles	The fleshy appendages that hang from each side of a chicken's beak.

Acknowledgements

The publisher wishes to thank the copyright holders for permission to use the following material:

Nipple water system design (page 58)
 – NSW Department of Primary Industries

'Perfect for bantams' house design (pages 60–63)
 – Dick Staie and Terry Towe, Morristown, New York, US

'Perfect for standards' house design (pages 64–65)
 – NSW Department of Primary Industries

2.4 m × 2.4 m layer house design (pages 66–67)
 – John Skinner, University of Wisconsin

6 m × 6 m layer house design (adapted, pages 68–69)
 – John Skinner, University of Wisconsin

Kind thanks also to veterinarian Patricia Macwhirter, of the Burwood Bird and Animal Hospital, Victoria, for her expert advice.

Index

albumen (egg white) 22
anatomy and physiology 17–23
Ancona breed 28, 40, 117
Andalusian breed 28, 40, 117
Araucana breed (Easter Egg Chickens) 28
artificial incubators see incubators
artificial lighting 59, 81, 84
aspergillosis (brooder pneumonia) 90
Australian Poultry Standards 40, 115
Australorp breed 28–9, 39, 40, 117
avian influenza (bird flu) 90–1

bantams 9, 38, 39, 40, 44
   house designs for 60–3
   perches for 53, 54
   size 27
   space needed 37–8
Barbu d'Anver breed (Belgian Bantam) 29, 40, 117
Barbu d'Everberg breed 29
Barbu d'Uccle breed 29
Barbu de Watermaal breed 29
Barnevelder breed 29, 40, 117
beak 18, 99
   trimming 18, 19, 101
behaviour and psychology 23–6
Belgian Bantam (Barbu d'Anver) 29, 40, 117
bird flu (avian influenza) 90–1
blood spots in eggs 86
botulism (limber-neck) 79, 91
Brahma breed 29, 39, 40, 117
breeds 27–37
   websites with images 130
breeders (contact details) 123
breeding
   reasons for 11
   setting and hatching 103–8
   from show chickens 116
breeding season 103
brooder (cage or box) 111–13
brooder (hen) 104–6
brooder pneumonia (aspergillosis) 90
broodiness 25–6
bumblefoot 91–2
buying chickens 14, 73–6, 86–7, 115–16

calcium deficiency 94, 99
Campine breed 30, 40, 117
candling 108
cannibalism 24–5, 38, 46, 58, 97, 99, 101
categorising chickens 27
chickens
   breeds 27–37, 130
   buying 14, 73–5, 86–7
   catching and handling 82
   for children 3–4, 40
   do they have teeth? 14
   easily bored 58
   for egg-laying 39
   flying 9
   how many to keep? 13, 37–8
   lifespan 14
   as pets 5, 12–13
   showing 4, 27, 39–40, 101, 115–19
   ten steps to keeping 15
   time commitment to 38
   twenty common questions about 7–14
   why choose chickens? 2–5
   see also chicks; hens; pullets
chicks
   buying 73–4, 86–7

first six weeks 111
hatching 107
incubator-hatched 110
pasting up 112
sexing 113
children, chickens for 3–4, 40
chronic respiratory disease (CRD) 92
classes of chicken 27
cleaning tasks 8, 80
cloaca 20, 21, 22
clubs and associations 122–3
coccidiosis 92–3
Cochin breed 30, 39
cold weather 81, 95
  breeds for 39
collecting eggs 11, 38, 84–5
comb and wattles 18–19, 95
commercial feed 77–8, 79
coop 41
  building 42–6
  cleaning 8, 80
  climate control 45–6
  considering neighbours 43–4
  mobile 43, 44–5
  smell 8, 44
  space per chicken 46
Cornish (Indian Game) breed 32, 40, 117
costs 2
council regulations 7, 9–10, 15, 37, 43
CRD (chronic respiratory disease) 92
crop 20, 21
crop bound 93
crossbreeds 27

deep-litter system 49
digestive system 20–1
diseases 86–98, 116
  epidemics 98–9
  vaccinations 87, 96, 97
Dorking breed 30, 40
dropping pit beneath roost 54
dust baths 26

earlobes 19
Easter Egg Chickens (Araucana breed) 28
eggs 2, 4
  average daily production 38, 59
  bantam 39
  blood spots in 86
  choosing breeds for 39
  cleaning 85
  collecting 11, 38, 55, 84–5
  colour 19, 39
  fertilisation 22
  freezing 85–6
  incubation period if fertilised 25
  internal structure 83
  production process 21–3
  rooster not needed 7, 24
  for setting and hatching 103–4
  storing 12, 85–6
  testing freshness 12
  see also laying
egg bound 93–4
egg cartons, reusing 86
egg-eating behaviour 99
egg shell 22–3, 78, 85, 99
egg-turning by brooder 106
egg yolk related peritonitis 94–5
epidemics 98–9
escape prevention 9, 71, 100–101
eyes 19

Faverolle breed 30–1, 40, 117
feathers 20, 25
feather pecking 99–100
feed and equipment suppliers 124–9
feeders 57–8
feeding 13, 23–4
  calcium for layers 78
  grinding eggshells 99
  grit 78
  how much do they eat? 77
  kitchen scraps 77, 78–9

feeding (continued)
    mash, crumble and pellets 77–8
    meat 79
    nothing rotten 79, 91
    organic feed 78
    scratch 79
fertilised eggs
    buying 73
    incubation period 25
fleas 87–9
fowl pox 87, 95
free-ranging
    in backyard 9, 79
    inside the house 12
freezing eggs 85–6
Frizzle breed 31, 40, 117
frozen combs 95

gizzard 20, 21
grit for digestion 78

hairworms 89
Hamburg breed 31, 40, 117
handling chickens 82
hatching time 110
hay (avoiding) 55, 90
health problems
    diseases 86–98, 116
    see also calcium deficiency; crop bound;
        egg bound; feather pecking; frozen
        combs; injured birds; mites, lice and
        fleas; pasting up; prolapse; worms
hens 7
    brooders 104–6
    buying 73–5
    free-ranging 9, 79
    noisiness 8
    replacing primary layers 103
hot weather
    breeds for 38–9
    precautions 80–1
Houdan breed 31, 39

house 41, 42, 46–7
    artificial lighting 81, 84
    cleaning and maintenance 80
    doors for people and chickens 49–50
    electricity supply 59
    feeders and waterers 57–8
    floor 48–9
    frame 47–8
    heating 59
    litter for floor 49
    nesting boxes 55–6
    ramp 49
    roof 51–2
    roost 52–4
    two compartments 47–8, 49
    windows and ventilation 50–1
house designs 60–9
hygiene 8, 80
    see also health

incubators
    candling after a week 108
    fertilised eggs for 103
    forced air and still air 108–9
    hatching time 110
    placing eggs 109–10
    reasons to use 108
    turning eggs 110
incubation period 106–7
Indian Game breed (Cornish) 32, 40, 117
infectious bronchitis 87, 95–6
infectious coryza 96
infectious laryngotracheitis 87, 96
injured birds 25

Japanese Bantam breed 32
Jersey Giants breed 32–3, 39

kitchen scraps 13, 76, 77–8, 91

Langshan breed 33, 40, 118
large intestine 20, 21

laying
    artificial lighting increases 59, 84
    calcium needed 78
    fake eggs to promote 55, 84
    first egg of pullet 83–4
    first year most productive 11, 74
    and hours of sunlight 11, 59, 84
    how long will it go on? 11
    how often? 11, 38
    during moulting 11, 25
    using nesting box 84
    when does it start? 11
    world record 29
Leghorn breed 33, 39, 40, 105, 118
legs and feet 20, 24
lice 87–9
lifespan 14
limber-neck (botulism) 79, 91
litter 9, 49, 80

manure as fertiliser 4
Marek's disease 87, 96–7
mating process 103
Minorca breed 33, 40, 118
mites 26, 80, 87–9
moulting 11, 25

neighbours, considering 10, 15, 43–4
nesting boxes 41
    access from outside 45, 54, 55, 56
    no hay 55, 90
    one per four or five hens 54–5
    positioning 55
    sloping roof 55
    starting to use 84
    tunnel nest 56
New Hampshire breed 34, 39, 40, 118
Newcastle disease 87, 97
noise 7, 8
nostrils 18

oesophagus 20, 21

Old English Game breed 34, 40, 118
organic feed 77
Orpington breed 34, 40, 105, 118
ovaries 21, 22
oviducts 21, 22

pasting up 112
pecking order 10, 23–4, 53–4
Pekin breed 34, 40, 118
perches *see* roost
permit for chickens 9–10
pets, chickens as 5, 12–13
Plymouth Rock breed 34–5, 40, 118
point-of-lay pullets 74
Polish breed 35, 40, 118
poultry breeders 14
poultry clubs 14
predators, protection from 9, 49, 50, 70
preening 26
prolapse 97
pullets (young hens)
    buying 73
    point-of-lay 74
    preparing to lay 83–4
    six-week-old 74
    starting to lay 11
    vaccinated 74, 87
purebreeds 27

rats 44, 48, 51, 62, 106
reproductive system 21–3
Rhode Island Red breed 32, 34, 35, 39, 40, 118
roost 41
    dropping pit beneath 54
    ladder 52–3
    moveable 54
    perch circumference 53
    perch heights 24, 53, 54
    positioning 52, 54
    staggered 53–4
    swinging 53

roosters
- aggressiveness 7, 24
- clipping claws 114
- comb 19
- council regulations 7
- damaging hens during mating 114
- noise problem 7, 8, 24
- not needed to produce eggs 7, 24
- one per ten hens 103, 113
- protecting their hens and chicks 24
- spurs 24
- top of the pecking order 23

roundworms 89

run 41, 42
- dirt for dirt baths 70
- netting beneath the ground 71
- shaded area 70–1
- size 70
- switching runs 71

scaly leg mites 87
scratching 26
Sebright breed 35, 40, 118
setting and hatching
- breeding season 103
- candling 108
- egg selection 103–4
- egg storage 104
- egg-turning 106
- incubation period 106–7
- making a nest 105–6
- mating process 103
- selecting brooder 104–5
- *see also* incubators

sexing chickens 113
showing chickens 4
- Australian Poultry Standards 40, 115
- breeding from winners 116
- breeds for 27, 39–40
- buying stock to show 115–16
- categories (Sydney Show) 117–18
- and disease spread 116
- inbreeding problems 116
- no clipped wings 101

Silkie breed 35–6, 40, 104–5, 119
size and pecking order 23
small intestine 20, 21
smells 8, 44
space per chicken 37–8, 46
standards (chickens) 27
state poultry associations 121
state and territory govt departments 120–21
stickfast fleas 88
storing eggs 12, 85–6
straight-run chicks 73–4
Sussex breed 36, 105, 119

tapeworms 89
toes 20
Transylvanian Naked Neck breed 36
trimming beak 18, 19, 101
tuberculosis 98

uterus 21, 22

vaccinations 74, 76, 87, 95, 96, 97
- doing it yourself 87

vagina 21, 22
vent 20, 21, 22

washing chickens 13–14
water sources 24, 79
waterers 57–8, 63, 79
wattles 18–19
websites with images of breeds 130
Welsummer breed 36, 39, 40, 119
wing clipping 100–1
worms (intestinal) 89–90
Wyandotte breed 37, 40, 119

yolk formation 22